Creating Selective Overmatch

An Approach to Developing Cyberspace Options to Sustain U.S. Primacy Against Revisionist Powers

TOM WINGFIELD, QUENTIN E. HODGSON, LEV NAVARRE CHAO, BRYCE DOWNING, JEFFREY ENGSTROM, DEREK GROSSMAN, CHAD HEITZENRATER, RYAN JOHNSON, CHRISTOPHER PAUL, CLINT REACH, MICHAEL SCHWILLE, JOHN YURCHAK

Prepared for the U.S. Cyber Command
Approved for public release; distribution unlimited

For more information on this publication, visit **www.rand.org/t/RRA1943-1**.

Published by the RAND Corporation, Santa Monica, Calif.

Library of Congress Cataloging-in-Publication Data is available for this publication.

ISBN: 978-1-9774-1154-9

About This Report

This report presents analysis examining how the United States should posture its military cyberspace forces in the cyber domain to address the strategic challenge posed by China. It represents a "clean sheet" approach to this problem that goes beyond building on current thinking and capabilities. The authors used the concept of "selective overmatch" (creating an overwhelming advantage over an adversary within a specific context) to develop a framework for identifying "influence points"—elements that are indispensable to regime survival. The analysis then identifies a set of influence points for both China and the United States.

The report is intended to help U.S. Cyber Command (CYBERCOM) and other cyber domain actors in the U.S. military response to the 2022 National Defense Strategy's call for "integrated deterrence": working across warfighting domains, theaters, the spectrum of conflict, instruments of national power, and the U.S. network of alliances and partnerships. In particular, it is intended to inform CYBERCOM force employment in the near term and force planning and experimentation in the medium term.

The research reported here was completed in March 2023 and underwent security review with the sponsor and the Defense Office of Prepublication and Security Review before public release.

RAND National Security Division

This research was sponsored by U.S. CYBERCOM and conducted within the International Security and Defense Policy Program of the RAND National Security Research Division (NSRD), which operates the National Defense Research Institute (NDRI), a federally funded research and development center sponsored by the Office of the Secretary of Defense, the Joint Staff, the Unified Combatant Commands, the Navy, the Marine Corps, the defense agencies, and the defense intelligence enterprise.

For more information on the RAND International Security and Defense Policy Program, see www.rand.org/nsrd/isdp or contact the director (contact information is provided on the webpage).

Acknowledgments

We would like to thank Michael Clark of CYBERCOM for initiating this project, Bob Leverton for his guidance and engagement throughout the project, and Matthew Shellem for his supervision of the project. Our thanks to Rich Girven and Carl "Cj" Horn for their helpful reviews and comments. David Adamson provided valuable contributions to improve readability and structure, and Susan Arick also helped with refining the report. Kate Giglio's help with developing the figures was invaluable.

Summary

This report develops an approach to cyber operations to support the 2022 National Defense Strategy's call for "integrated deterrence" by examining how the United States should posture its military cyberspace forces in the cyber domain to address the strategic challenge posed by China. Integrated deterrence is defined in the strategy as "working seamlessly across warfighting domains, theaters, the spectrum of conflict, all instruments of U.S. national power, and the U.S. network of alliances and partnerships. Tailored to specific circumstances, the concept applies a coordinated, multifaceted approach to reducing competitors' perceptions of the net benefits of aggression relative to restraints."[1] Providing cyberspace options that integrate with the other warfighting domains represents a departure from the current U.S. Cyber Command (CYBERCOM) focus on technical operations in the cyber domain: defensive cyberspace operations on Department of Defense (DoD) networks, cooperative hunt forward missions on allied and partner networks, and offensive cyberspace operations against adversary networks. Successfully contributing to an integrated deterrence strategy will mean evolving and expanding the conception of cyberspace operations and developing new concepts for their employment.

Creating Selective Overmatch in the Cyber Domain

Our approach is rooted in the concept of selective overmatch: creating an advantage over an adversary in conflict or competition by targeting "influence points"—elements of a country's political, economic, or societal strength—to achieve effects. Achieving selective overmatch first requires a holistic analysis of the adversary to identify its influence points, independent of the means used to try to affect those points. Only then can we examine how cyberspace capabilities can contribute to affecting those points as part of a broader strategy. Influence points at the policy level are elements

---

[1] U.S. Department of Defense, *2022 National Defense Strategy of the United States of America*, October 27, 2022, p. 1.

that are indispensable to national (in the case of the United States) or regime survival (in the case of the Chinese Communist Party). Policy influence points, in turn, are supported by political, military, diplomatic, and economic "strategic influence points" and by critical military, intelligence, and security capabilities—or "operational influence points."

We posited four postures toward conflict and competition that provide a basis for developing an array of options to achieve objectives: minimalist competition, maximalist competition, minimalist conflict, and maximalist conflict, which we refer to as the MC4 construct (Figure S.1). The minimalist end of the spectrum captures options that are optimized to be measured—largely defensive measures that are less provocative, and less escalatory actions that are intended to provide effects against the largest number of operational requirements with the minimum number of forces. The "maximalist" end, on the other hand, provides complementary options—more active, offensive, potentially more provocative, and escalatory

FIGURE S.1
The MC4 Construct: Postures in Competition and Conflict

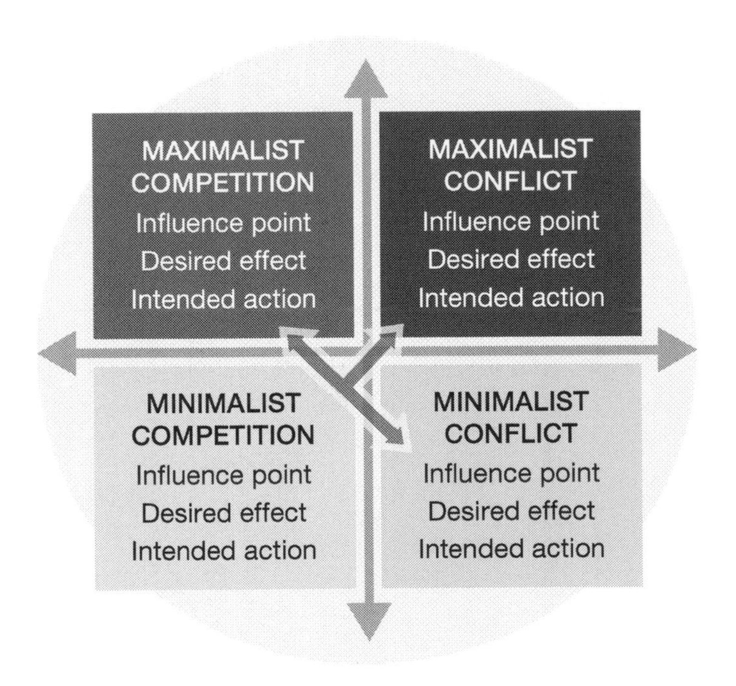

that require a greater military and political commitment to deliver effects against higher-priority influence points. Each posture frames a different set of tools, actions, and restraints against any particular influence point.

Selective overmatch relies on developing and evolving cyberspace operations across four broad areas of new capabilities needed to permit a single cyber commander to undertake all intended cyber action to produce all desired effects against all influence points.

Integrated cyber incorporates adjacent capabilities (cyber-cognitive in addition to cyber-technical organic intelligence and certain information-related and electronic warfare capabilities).

Integral cyber recognizes that any significant military operation in the 21st century is unthinkable without cyber components (e.g., encrypted communication; position, navigation, and timing; sensor integration; multi-platform targeting) that enhance speed, precision, and lethality.

Expanded cyber expands beyond traditional peacetime constraints in the minimalist-competition quadrant to include more assertive action in the maximalist-competition quadrant. Operations will also require strategic and operational agility to move from quadrant to quadrant more quickly than the adversary. This will entail developing and having options aligned to each of the four quadrants available to execute at any given moment.

Extended cyber broadens CYBERCOM's operational horizons across another dimension—beyond the active technical networks that have been the focus of operations up to this point to include visibility across supply chains, life cycles, and emergent networks.

We refer to these capabilities by the shorthand "I2E2."

Chinese Influence Points

The team identified the following Chinese influence points, based on China's priority policy-level goals:

- **Maintaining CCP political control.** Maintaining exclusive Chinese Communist Party (CCP) control is paramount to enacting all other policies. Since its founding by Mao Zedong, the People's Republic of China has refused to allow any viable political opposition in the system.

- **Ensuring domestic stability.** The Chinese government views economic development as the key to social stability.
- **Advancing domestic industry, science, and technology.** These advancements are seen as key to continued economic growth and domestic stability.
- **Building a global mercantile network for secure access to foreign critical materials.** Chinese organizations would ensure the establishment, operation, and security of such a hub-and-spoke series of bilateral trade with easily managed partners.
- **Establishing an Indo-Pacific order.** China aspires to create a Sinocentric order in the Indo-Pacific region that would be friendly to Chinese economic, security, and political interests and establish a dominant position against competitor nations.
- **Developing and maintaining a military capable of defending and promoting all these policy goals.** China continues to pursue ambitious military development to expand its capabilities to support foreign and domestic goals.

China's strategic goals align closely with these overarching policy objectives, and its operational goals are designed to advance toward these goals.

U.S. Influence Points

The United States has its own influence points that it must protect and defend at the policy, strategic, and operational levels. Influence points at the policy level are synonymous with enduring national interests. We identified four U.S. policy-level influence points.

Security of the United States. To secure its territory, protect against external threats, and sustain the United States as a state, the United States must provide for the common defense against external threats and ensure domestic tranquility by addressing internal threats. An adversary could seek to pose a threat from abroad to the United States' territory through physical attack or by undermining and exploiting fissures in domestic stability.

Prosperity. The United States derives its power through its economic might, which depends on robust economic activity domestically and access to world markets. An adversary could seek to undermine the United States economically.

Liberty. The United States values individual liberty to pursue social and economic well-being, supported by a constitutional government that protects minority rights and mediates disputes through judicial and political processes rather than violence. An adversary could seek to undermine this social compact to sow further division in the United States and convince significant numbers of Americans that violence or other nondemocratic means are legitimate ways to act.

Extended Values. The United States sees democratic governance and individual liberty as universal values that support security and prosperity, domestically and abroad. An adversary could seek to undermine the acceptance of these values internationally or present an alternative set of values, as the Soviet Union did during the Cold War.

These influence points represent enduring U.S. national interests. U.S. strategic goals offer ways to pursue these policy objectives, and operational goals are designed to defend them. However, there is no clear demarcation between the defense of these influence points and the actions that will be required to protect them. Just as the protection of any one influence point will require a carefully orchestrated set of actions, the complexity of the U.S. legal, political, and economic systems means that some individual actions will affect multiple influence points simultaneously.

In addition, U.S. influence points exist in a global ecosystem; there is no clear demarcation between domestic and foreign interests. One of the distinguishing asymmetries between the United States and the People's Republic of China is that most U.S. international interests are non-zero-sum in that increasing security, liberty, and prosperity overseas enhances the same interests in the United States.[2] Chinese international interests, on the other hand, are mostly zero-sum, where Chinese political, economic, or military gains are at the expense of adversary or even client states.

---

[2] We acknowledge, however, that not all segments of U.S. society benefit equally; indeed, some will experience negative consequences.

Findings

- Defending U.S. influence points is just as important as holding Chinese influence points at risk.
- Cyberspace operations are likely to be more appropriate and impactful for some influence points compared with kinetic operations.
- Influence points must be defended, or held at risk, in both competition and conflict.
- Operations against influence points are nonlinear and involve multiple causes and effects; affecting a single influence point may require multiple diverse actions while a single action may affect multiple influence points.
- Changes in some influence points will create changes in others; there will be a need to monitor the complex military and political ecosystem in which they exist.
- Influence points are not passive targets but rather part of a complex adaptive system; the adversary continuously observes, orients, decides, and adapts.

Recommendations

Near Term

Selective overmatch and the I2E2 set of capabilities comprise a concept intended to address a need for a broader and more comprehensive way to address the challenge of revisionist powers such as China. The concept has potentially far-reaching implications for doctrine, plans, organizational relationships, force generation and employment, and capabilities. Before advancing to these stages, however, CYBERCOM will need to test the hypothesis to determine how well it meets the stated need and develop implementation plans.

Recommendation 1: CYBERCOM should test the concept in wargames and tabletop exercises to evaluate the degree to which it meets strategic and operational objectives.

Medium Term

After a wargame series, experimentation with existing capabilities and teams can demonstrate how to make the concept "real" in an operational environment.

Recommendation 2: Using units from the Cyber National Mission Force and in partnership with the services, CYBERCOM should develop an experimentation plan. The purpose is to develop a replicable model for creating additional integrated cyber teams to bring these enhanced capabilities to bear in all four cyber operational postures against all assigned influence points.

Long Term

Once CYBERCOM and other components of the Department of Defense have tested the concept and meta-capabilities and determined they make sense in an operational setting, adopting and employing them at scale will require changes across the DOTMLPF (doctrine, organization, training, materiel, leadership and education, personnel, and facilities) spectrum.

Recommendation 3: Contingent on a determination that the concept and meta-capabilities are necessary and effective, CYBERCOM should work with the other combatant commands (CCMDs), the services, the Joint Staff, and the Office of the Secretary of Defense to develop requirements for a revised force structure to organize, train, equip, and present integrated cyberspace forces to CYBERCOM.

Recommendation 4: Operational planning at CCMDs would benefit from better integration of cyberspace expertise from the start of the planning process. Although cyberspace operations are included (typically in the plan's operations annex as an appendix), the planning process still primarily focuses on operations in the physical domains as the primary efforts and operations in other domains (e.g., cyber, information) as secondary or supporting efforts. This integration will require CYBERCOM to work with the Joint Staff to revise the approach to planning, which often addresses domains in distinct silos.

Recommendation 5: This prior set of recommended changes will call for revisions in operational relationships not only at the CCMD-to-CCMD level, as CYBERCOM may need operational control over a more diverse set of forces, but also at subordinate levels. These changes will need to be explored and refined as part of the roadmap from wargame to experiments to implementation. Changes to force allocation and assignment will require Secretary of Defense approval and, should the mission set of CYBERCOM expand to include additional cyber-adjacent capabilities, may also require an update to the Unified Command Plan and approval by the Commander in Chief.

Contents

Figures

CHAPTER ONE

Introduction

Since its creation in 2009, U.S. Cyber Command (CYBERCOM) has engaged in operations to defend the United States against cyberattacks, operate and defend the Department of Defense Information Network (DoDIN), integrate cyberspace operations into combatant command (CCMD) plans, protect U.S. elections, and support allies and partners in defending themselves from malicious cyber activity. Russia's invasion of Ukraine in February 2022 caused the command to increase its operational tempo, which CYBERCOM commander General Paul Nakasone acknowledged was "already high."[1] As a CCMD with global responsibilities, CYBERCOM is tasked with providing support to regionally oriented CCMDs, operating and defending DoDIN, and defending the nation in cyberspace—a daunting set of missions that compete for highly skilled forces and employment of sophisticated capabilities.

The 2022 National Defense Strategy poses new challenges for CYBERCOM. It introduces the broad concept of "integrated deterrence."[2] This strategic

[1] Paul M. Nakasone, "Posture Statement of Gen. Paul M. Nakasone, Commander, U.S. Cyber Command Before the 117th Congress," U.S. Cyber Command, press release, April 5, 2022.

[2] U.S. Department of Defense, *2022 National Defense Strategy of the United States of America*, October 27, 2022, p. 22. The elements of "integrated deterrence" are integration across domains, regions, the spectrum of conflict, the U.S. government, and allies and partners. There are numerous points of contact between the integrated deterrence strategic concept and the integrated, integral, extended, expanded (I2E2) construct that is developed in this report. Within I2E2, integral cyber addresses integration across domains; expanded cyber calls for greater integration across the spectrum of conflict; and extended cyber involves closer integration across the U.S. government, regions, and allies and partner nations.

concept "entails working seamlessly across warfighting domains, theaters, the spectrum of conflict, all instruments of U.S. national power, and the U.S. network of alliances and partnerships. Tailored to specific circumstances, the concept applies a coordinated, multifaceted approach to reducing competitors' perceptions of the net benefits of aggression relative to restraints."[3] Additionally, the 2023 *National Cybersecurity Strategy* calls for integrated federal efforts to disrupt malicious cyber activity, including by nation-states; and the 2022 *National Security Strategy* notes that the United States is "in the midst of a strategic competition to shape the future of the international order," underscoring the need for new concepts and capabilities to address these challenges.[4]

What Does Integrated Deterrence Mean for CYBERCOM?

To help answer this question, CYBERCOM turned to RAND's National Defense Research Institute (NDRI). NDRI was asked to determine how CYBERCOM can support the new National Defense Strategy and to take a "clean sheet" look at how CYBERCOM can best address the expanding and competing demands on its capabilities. In particular, CYBERCOM asked NDRI to explore concepts for integrating deterrence and sustaining United States' primacy *against the challenge of revisionist powers* (i.e., those powers acting to weaken or replace the existing hegemon, in this case, China seeking to displace the United States in its global leadership role).[5]

---

[3] DoD, October 27, 2022, p. 1.

[4] Joseph R. Biden, *National Cybersecurity Strategy*, Washington, D.C.: White House, March 2023; and Joseph R. Biden, *National Security Strategy*, Washington, D.C.: White House, October 2022.

[5] China and Russia are discussed as revisionist powers in Donald J. Trump, *National Security Strategy of the United States of America*, Washington, D.C.: White House, December 2017, p. 25. For a historical view on revisionist powers and great power rivalry, see Michael J. Mazarr, Samuel Charap, Abigail Casey, Irina A. Chindea, Christian Curriden, Alyssa Demus, Bryan Frederick, Arthur Chan, John P. Godges, Eugeniu Han, et al., *Stabilizing Great-Power Rivalries*, Santa Monica, Calif.: RAND Corporation, RRA456-1, 2021.

In response, our analysis addresses an overarching research question: How should the United States posture its military cyberspace forces in the cyberspace domain to address the strategic challenge posed by China?

The analysis consists of four tasks:

1. To provide analytic background, we summarize key theories of international relations to help identify potential opportunities and risks in the U.S. approach. The three primary schools of thought in international relations theory—*realism*, *liberalism*, and *constructivism*—do not agree on the primary motivations and dynamics of the international system, and they have different views on the interplay between a rising power and a status quo power. We explore these different schools to provide a range of views on the U.S.-China competition and draw implications for our subsequent analytic insights.
2. To provide additional background, we review Department of Defense (DoD) cyberspace strategy, concepts, doctrine, and plans as a baseline for understanding new cyber domain concepts.
3. We develop a framework for thinking about how to generate cyberspace options in competition and conflict. In this framework, the current and largely network-based approach to cyberspace operations evolves to include more expansive ideas of what cyberspace operations should encompass. The framework applies the new concept of "selective overmatch" to the competition (and potential conflict) with China. Selective overmatch acknowledges that the United States cannot maintain superiority in all domains at all times and must selectively create advantage in the areas that will most likely contribute to achieving U.S. objectives.
4. We apply the framework to identify and assess vulnerabilities in the Chinese and U.S. cyber domains. We first evaluate China's national objectives and derive what we term "influence points" at the policy, strategic, and operational levels. This analysis provides the starting point for understanding where and how the United States could apply cyberspace capabilities to achieve its objectives. We next identify U.S. strategic objectives and its own influence points that could, if targeted effectively, allow the Chinese (or other adversaries) to prevent the United States from achieving its national goals or sustaining the international order.

The purpose of our analysis is to inform CYBERCOM force employment in the near term and force planning and experimentation in the medium term.

Research Limitations

A note on limitations is necessary. We conducted this research and analysis at the unclassified level to allow wider dissemination and discussion of our findings. Our sponsor requested analysis that would support a broader engagement with stakeholders in the executive branch, in Congress, and with the public about the requirements for CYBERCOM and DoD as a whole to evolve the concept of cyberspace operations and educate all stakeholders on the limitations and opportunities cyberspace presents. This approach limits what we can say about U.S. and Chinese capabilities, weaknesses, and opportunities. Nevertheless, we believe these limitations do not undermine the overall analysis and conclusions we present. The reader should not, however, draw any conclusions from this report about current or future U.S. capabilities, plans, or intentions beyond what is available to the general public. Additionally, we examine one dyad in this report—the U.S.-China competition—and not the interplay across other dyads, such as the U.S.-Russia competition. To address the potential demands the concept of selective overmatch and its supporting capabilities would place on CYBERCOM, we focus on the United States and do not address how to integrate its allies and partners into the framework we lay out.

CHAPTER TWO

The Application of International Relations Theory to Competition and Conflict in Cyberspace

To begin exploring the dynamics between the People's Republic of China (PRC) and the United States and what those dynamics might mean for the application of cyberspace operations to shape, influence, and respond to the PRC, we draw on international relations theory. We use this discipline to describe the existing relationship between the United States and China and to inform potential ways and means cyberspace operations can be used to affect the power dynamic between the two countries. We provide an overview of the three prominent schools of international relations theory (realism, liberalism, and constructivism), analyzing how effectively the primary schools of international relations explain conflict and competition dynamics in cyberspace, and finally determining whether the cyber domain can be described by existing international relations theories.

Overview of Three Prominent Schools of International Relations

International relations theory as a discipline attempts to provide causal explanations for why states and actors behave the way they do. Within the field, there are three prominent schools—realism, liberalism, and constructivism—that each emphasize varying factors as to why states engage in certain action and at certain periods of time. Because of the complexity of international affairs, these theories provide mental schemas that seek to identify critical elements in state behavior by relying on varying units of analysis and points of emphasis.

Realism

Perhaps the most well-known and certainly the most influential theory in foreign policy for the past century, realism characterizes state behavior as self-interested and confrontational. In terms of armed conflict, realist theory views conflict as inevitable. *Anarchy* is a central tenet, generally defined as the condition in the international system characterized by the absence of a superior power regulating the interactions of states. Because states are the highest element within the international system, they act independently.

> Anarchy is the condition in the international system characterized by the absence of a superior power regulating the interactions of states. The anarchic system does not mean there is a complete lack of order. Rather, because states are the highest unit of agency, they act to serve their own interests (Lechner, 2017).

In addition to anarchy, the realist perception relies on the following assumptions:

- All great powers possess the power to harm others.
- States are uncertain whether other states will refrain from exerting that power.
- States seek to maintain their survival and maximize their interests at each other's expense.
- States are rational actors.

Classical Realism

Classical realism, which was first defined by the political scientist Hans Morgenthau during the early years of the Cold War, has become a mainstay in international politics.[1] Influenced by Thomas Hobbes's *The Leviathan*,

[1] Realist literature is extensive, and we do not purport to present a complete picture of it and its many variations. Seminal works are referenced. Others include Robert Jervis,

classical realists arguably take the bleakest view of state behavior in arguing that, like humans, states possess a natural desire to maximize their absolute power in order to exert control over others. For classical realists, this manifests in what is now typically referred to as "hard power," or as Morgenthau defined it: "material strength, especially of a military nature."[2]

Defensive Realism Holds That Nations Seek Only a Balance of Power

Kenneth Waltz developed defensive realism, arguing that states only seek to maintain the current balance of power. Viewing the world from what he termed the "third image" (the international arena is made up of various states), Waltz believes that the best way for states to preserve their security in an anarchical world is by maintaining the status quo.[3] A key concept in this school is the "balance of power." By seeking to become a global hegemon as stated by Mearsheimer and Morgenthau, states draw attention to themselves and appear aggressive. This will lead other states to balance that initiative and put the aggressive state in danger. It is for this reason, Waltz argues, that states should only seek to maintain the status quo and balance threats when external events occur. This can be seen in the way many states act because few aspire to be the global hegemon. Instead, they seek to maintain the status quo and win marginal victories when they can.

"Realism in the Study of World Politics," *International Organization*, Vol. 52, No. 4, Autumn 1998, pp. 971–991; John J. Mearsheimer, "Anarchy and the Struggle for Power," in *The Tragedy of Great Power Politics*, New York: W. W. Norton, 2001; and Hans J. Morgenthau, *Politics Among Nations: The Struggle for Power and Peace*, 2nd ed., New York: Knopf, 1956.

[2] Morgenthau, 1956, p. 186.

[3] Waltz describes three "images" in his classic work *Man, The State, and War*. The first is international conflict and human behavior, which claims that conflict originates in motivations inherent to human behavior, like selfishness. The second image claims that the first image is necessary but not sufficient to explain conflict and that "the international organization of states is the key to understanding war and peace" (Kenneth Waltz, *Man, The State, and War*, New York: Columbia University Press, 1959, p. 81). The third image addresses the role of anarchy in the international system.

Liberalism

Like realism, the liberal school also believes that states exist in a constant state of anarchy.[4] However, liberalism diverges from realism in that liberals believe states are not forced to exist in a self-help and zero-sum game to secure their own power. They can instead develop mechanisms to avert anarchy and develop cooperative systems without conflict. Within the liberal school, there are several notable theories. Most relevant of these are neoliberal institutionalism and economic interdependence. In viewing the world through the lenses of these two camps, liberals see state conflict as unnecessary if managed through the proper channels.

Neoliberal Institutionalism: States Can Subvert the Zero-Sum System Through Institutions

In contrast to the realist school, the neoliberal institutionalist theory presumes that states can actually subvert the zero-sum, self-help system by developing institutions that lead to shared expectations and common rules to the game. As Robert Keohane discusses, institutions lead to more transparency and credibility, thereby allowing states to trust each other.[5] This in turn means that states see state actions not as confrontational but cooperative, because these institutions create a vehicle for states to communicate, share information, and lessen the doubt caused by uncertainty in an anarchic system.

Economic Interdependence

Similar to neoliberal institutionalism, economic interdependence argues that conflict can be averted and the dynamics of anarchy overcome. Unlike

---

[4] The terms "liberalism" and "liberal" here are used in the classic political science sense of the words, not the more common use of the term in public discourse in the United States. Some of the seminal texts include Robert O. Keohane and Joseph S. Nye, Jr., *Power and Interdependence*, 4th ed., New York: Pearson Longman, 2012; John Gerard Ruggie, "International Regimes, Transactions, and Change: Embedded Liberalism in the Postwar Economic Order," *International Organization*, Vol. 36, No. 2, Spring 1982, pp. 379–415; and Robert O. Keohane, "The Demand for International Regimes," *International Organization*, Vol. 26, No. 2, Spring 1982, pp. 325–355.

[5] Robert Keohane, "International Institutions: Can Interdependence Work?" *Foreign Policy*, Spring 1998, pp. 82–96.

the institutionalists, economic interdependence theorists argue that this is achieved by developing cooperation through economic ties rather than institutions. Economic interdependence theory argues that by increasing trade that interweaves states' economic ties, these states are less likely to engage in armed conflict, since war disrupts trade and carries more outsized ramifications. As Dale Copeland famously put it, "States would rather trade than invade."[6]

Constructivism

Unlike the two previous camps discussed, realism and liberalism, the constructivist camp dismisses the idea of anarchy altogether. Adherents of this theory argue that anarchy, like other concepts such as norms, values, and state strength, is socially constructed and based on one's perception and understanding of the world. While there are varying theories within the constructivist camp, the most notable regarding the U.S.-China competition is the norms, values, and ideas model.

Norms, Values, and Ideas Shape Behavior: The Nuclear Taboo

Within the constructivist camp, the predominant theory emphasizes the culture and values of a state rather than the environment in which the state exists. Alexander Wendt, an influential scholar in the constructivist camp, states that "anarchy is what states make of it," meaning that states view the world based on their beliefs and cultures rather than the structural perspective the realist and liberal camps assume.[7] As such, state behavior and reactions to other states' behavior are in the eye of the beholder.

One such theory, as offered by Nina Tannenwald, is that throughout the Cold War and in the present, states have refrained from using nuclear weapons. This is because of the emergence of a "nuclear taboo" originating among states and a shared set of norms and values that prohibits the use

---

[6] Dale C. Copeland, "Economic Interdependence and War: A Theory of Trade Expectations," *International Security*, Vol. 20, No. 4, Spring 1996, pp. 5–41.

[7] Alexander Wendt, "Anarchy Is What States Make of It: The Social Construction of Power Politics," *International Organization*, Vol. 46, No. 2, Spring 1992, pp. 391–425. See also Ted Hopf, "The Promise of Constructivism in International Relations Theory," *International Security*, Vol. 23, No. 1, Summer 1998, pp. 171–200.

nuclear weapons. After World War II, states and societies informally and collectively agreed to not use nuclear weapons. She goes on to say that "the taboo is not the behavior (of nonuse) itself but rather the normative belief about the behavior."[8]

Clearly, from a military balance perspective, realists and liberals would expect a state living in the self-help system to arm itself with nuclear weapons to deter the prospect of an adversary using them. Through the lens of the constructivist argument, however, this is not the case. Because of the powerful influence of norms and culture, state behavior can be influenced by shared beliefs that can result in prohibitions on certain actions.

Theorizing Conflict in the Cyber Domain Through the Lens of International Relations Theory

What is unique about the cyber domain, and how do these theories apply? There is surprisingly little literature that tackles this question.[9] The first difference between the cyber domain and other domains is the logical rather than spatial connection between places. Within the cyber domain, distance between nodes is measured not as space between geographical points but in terms of their logical distance, which can be measured, for example, as a

---

[8] Nina Tannenwald, "Stigmatizing the Bomb: Origins of the Nuclear Taboo" *International Security*, Vol. 29, No. 4, Spring 2005, p. 8.

[9] Our literature review identified a handful of articles that address one or more of the international relations schools and their views on conflict and competition in cyberspace. These include Robert Reardon and Nazli Choucri, "The Role of Cyberspace in International Relations: A View of the Literature," paper prepared for the 2012 International Studies Association Annual Convention, San Diego, Calif., April 1, 2012; Rika Isnarti, "A Comparison of Neorealism, Liberalism, and Constructivism in Analysing Cyber War," *Andalas Journal of International Studies*, Vol. 5, No. 2, November 2016; Johan Eriksson and Giampiero Giacomello, "The Information Revolution, Security, and International Relations: (IR)relevant Theory?" *International Political Science Review*, Vol. 27, No. 3, July 2006), pp. 221–244; and Constantine J. Petallides, "Cyber Terrorism and IR Theory: Realism, Liberalism, and Constructivism in the New Security Threat," *Inquiries Journal/Student Pulse*, Vol. 4, No. 3, 2012. For a discussion of the limitations of prior theory that extends from realism and constructivism, see Michael P. Fischerkeller, Emily O. Goldman, and Richard J. Harknett, *Cyber Persistence Theory: Redefining National Security in Cyberspace*, New York: Oxford University Press, 2022.

number of hop points—all locations are within only a few logical connections of each other.[10]

Because of their focus on positive-sum interactions and cooperative game theory, liberals could see the distance-reducing effects of cyberspace as a method of reducing transaction costs—that is, reducing the price of transmitting information-based goods and services. According to existing literature, one of the key mechanisms through which international regimes persist is by sharing information and reducing transaction costs. Liberals might therefore see cyberspace as an area where increasing integration between nations (either naturally or through international organizations) could preserve the existing international regime, whereas revisionist powers may try to undermine this idea by restricting or entirely separating parts of the internet from each other.

Constructivists, on the other hand, might think of this connected domain as a new landscape shaped by cultural norms and initial technological restrictions and as a potential mechanism by which norms are homogenized across physical and national boundaries.

An oft-discussed aspect of the cyber domain is the difficulty of attributing actions to specific actors within it. The "attribution problem" is a result of the logical distances between locations mentioned above as well as the massive scale of the cyber domain that makes it difficult to identify exactly where any action originated. The inherent uncertainty in positively attributing an action to a political actor like a nation-state also has a nontechnical dimension; a claim of attribution by one state will often be denied by the other state, and the techniques for attribution are often debated across technical and political communities.[11] This makes the deterrent effect of retaliation much less salient, as attackers can be very difficult to identify and will often deny responsibility. Both liberals and realists see this as a destabilizing

---

[10] A hop point refers to each router that a packet of data passes through on the way to its destination; it is often used to count the logical distance that a packet has to traverse through a traceroute from origin to destination. The logical distance between points is not constant. Packets of information sent from one point to another can take multiple paths that change continuously.

[11] See Andrew Grotto, "Deconstructing Cyber Attribution: A Proposed Framework and Lexicon," *IEEE Security & Privacy*, Vol. 18, No. 1, January–February 2020, pp. 12–20.

influence. While realists are likely to perceive this as ultimately a zero-sum impact on the security environment, liberals are likely to see an opportunity for states to coordinate to resolve this problem, making the international environment safer for all countries. Constructivists, on the other hand, would be more interested in how this environment of anonymity is likely to shape the norms and culture of cyberspace. Specifically, they might compare it to the norms around other areas where attribution is especially difficult, such as espionage.

A specific characteristic of the cyber domain is the explicit set of offense-defense trade-offs that states face when engaging in offensive cyber operations. Cyberattacks typically rely on vulnerabilities in the software of the target,[12] since the underlying software that states, organizations, and individuals use is often similar, frequently containing the same or very similar vulnerabilities. Because of this, states must choose to identify and fix vulnerabilities that threaten both themselves and potential targets of cyber operations or to preserve these vulnerabilities for use in future cyber operations, simultaneously leaving themselves open to those same attacks.

In sum, the various schools of international relations have ideas about competition that apply to the cyber domain in the following ways:

- The realists perceive the cyber domain as one of competition and a new avenue of coercive force to use against adversaries.
- The liberals, like the realists, perceive the cyber domain as an extension of already-existing international dynamics. However, unlike the realists, liberals are much more likely to see the cyber domain as a new way to stabilize existing international dynamics though new mechanisms of information-sharing and transaction cost reduction as well as providing a unifying security objective for all states heavily involved in the domain.
- The constructivist perception of the cyber domain is fundamentally different from that of the liberals or realists. Constructivists see the cyber domain as a fundamentally different, human-constructed realm with a distinct culture and set of norms, which may not reflect the same norms as those in the physical domains.

[12] Hardware vulnerabilities can also be exploited.

In the context of the U.S.-China competition, there are a variety of ideas about how each country is likely to perceive the cyber domain.

Because of the destabilizing influence of the attribution problem, realists would likely say that the cyber domain presents new opportunities for revisionist powers to challenge the status quo power.[13] In essence, realists see the cyber domain as a new competition space where the power distribution is different from the physical domain and where the deterrent effect of retaliation is significantly reduced. *Accepting this assumption leads to the conclusion that China would vigorously pursue operational cyberspace capacity because it provides advantages to revisionist power to challenge the status quo power.* The United States, on the other hand, is likely to seek to preserve its existing hegemon status either through attempting to constrain Chinese power in the cyber domain in traditional ways or to work to improve attribution in the cyber domain to restore the traditional deterrent effect.

Liberals are likely to see the effects of engaging in the cyber domain in a way diametrically opposed to the realists. Liberals believe that the cyber domain provides opportunities for the existing status quo power to preserve the existing international regime.[14] The United States could either create or co-opt existing international institutions to help standardize, regulate, or coordinate actions in cyberspace. It would do this to better coordinate with its allies, as well as to help align the interests of various other countries with those of the United States. International law enforcement is a clear example. By extension, this would also assist in helping to attribute actions to actors

[13] In conducting our literature review, we identified a small number of articles that present an argument on conflict in cyberspace that is grounded explicitly in one or more of the three schools we examine here. Our presentation of how each school would present its view on conflict in cyberspace is therefore an extrapolation from the literature. For the realist school, see Anthony Craig and Brandon Valeriano, "Realism and Cyber Conflict: Security in the Digital Age," in Davide Orsi, J. R. Avgustin, and Max Nurnus, eds., *Realism in Practice: An Appraisal*, Bristol: E-International Relations Publishing, 2018; Marcio Rocha and Daniel Farias da Fonseca, "The Cyber Issue and Realist Thinking," *Revista da Escola de Guerra Naval (Rio de Janeiro)*, Vol. 25, No. 2, May–August 2019, pp. 517–543; Mary McEvoy Manjikian, "From Global Village to Virtual Battlespace: The Colonizing of the Internet and the Extension of Realpolitik," *International Studies Quarterly*, Vol. 54, No. 2, June 2010, pp. 381–401.

[14] Manjikian, 2010. See also Joseph S. Nye, Jr., *Cyber Power*, Cambridge, Mass: Belfer Center for Science and International Affairs, May 2010.

within cyberspace. China, on the other hand, would seek to either create a competing set of international organizations or segregate itself from the global internet through various means, while also encouraging other countries to do the same to deprive the United States of the desired effects of international cooperation within the cyber domain.

The constructivist view on conflict in the cyber domain is less clearly articulated, particularly in how it sees cyberspace's role in the competition between the United States and China.[15] In this view, the United States sees the cyber domain as a public space that should be governed by norms of behavior, while China sees cyberspace as potentially more threatening to its domestic stability as evidenced by things such as the Great Firewall (the Chinese Ministry of Public Security's system for monitoring and controlling access to information on the internet coming into and within China) and general censorship of information in cyberspace. In addition, cyberspace and the norms and culture that develop around it will influence policies and perceptions in other domains, just as the culture and norms of those domains can have an influence on cyberspace.

The theories of the three schools of international relations explored in this chapter have different perceptions of how the competition between the United States and China will play out in cyberspace. Although each school makes a good case for its view of the world, there is not a clear winner in terms of predictive power. Subscribing to a realist view would indicate the United States should expect conflict in cyberspace with China and should prepare to confront aggressive actions in cyberspace. On the other hand, a liberal view would indicate the United States is not destined to clash with China in cyberspace and can use existing, or create new, institutions and frameworks to govern behavior in cyberspace and increase Chinese confidence in its own security. Finally, a constructivist view is potentially more sympathetic to the liberal view in that the United States could seek to estab-

[15] There are not many explicitly constructivist articles on conflict in cyberspace, but several authors have addressed how establishing norms of state behavior can contribute to more stability. See, for example, Martha Finnemore and Duncan B. Hollis, "Constructing Norms for Global Cybersecurity," *The American Journal of International Law*, Vol. 110, No. 3, July 2016, pp. 425–479. For one discussion on the U.S.-China competition, see Benjamin C. Jamison, "A Constructivist Approach to a Rising China," *Journal of Indo-Pacific Affairs*, May 19, 2021.

lish norms of behavior in cyberspace that account for Chinese interests and views while protecting the United States' ability to preserve its security through cyberspace.

So what does this mean for how the United States *should* think about the application of power in cyberspace to achieve its goals? It would seem prudent to seek cooperation with China where possible, especially given the immense economic and growing military power it has, while understanding that conflict with China in cyberspace is probable, although not inevitable. As we will explore in later chapters, Chinese strategic interests are inextricably tied to the interests of the Chinese Communist Party (CCP) and regime survival. To the extent that the CCP sees the United States as a threat to those interests, China will want to reshape the international order and balance or confront the United States, including in cyberspace. Having options that prepare the United States for conflict while allowing the flexibility to de-escalate and even cooperate would provide a balanced portfolio of capabilities and options.

CHAPTER THREE

Current U.S. Approach to Cyberspace Operations

This chapter explores current U.S. cyber operations, with a brief discussion along three key dimensions: organization of cyber forces, including the service components of cyber forces; acquisition of cyber capabilities, which includes research, development, and infrastructure; and the employment of cyber forces as part of operations, inclusive of planning, training, and rehearsal. The discussions within each of these dimensions provide necessary background and frame the primary contribution of this chapter: the identification of observed and projected limitations that may arise in a direct conflict involving peer adversaries.

Organization of Cyber Forces

Leadership of U.S. CYBERCOM is provided by a four-star general or flag officer who maintains a "dual-hatted role" as commander of CYBERCOM and the director of the National Security Agency (NSA). The Cyber Mission Force (CMF) within CYBERCOM consists of multiple elements, each with its own purpose and structure. The **Cyber National Mission Force** (CNMF) is composed of national mission teams (NMTs) responsible for defeating significant threats to the DoDIN and the nation. The **Cyber Combat Mission Force** (CCMF) consists of numbered cyber (combat) mission teams (CMTs) and associated combat support teams. These teams conduct operations in support of the "missions, plans, and priorities of the geographic and functional combatant commanders (CCDRs)."[1] The **Cyber Protection Force** (CPF) consists of cyber protection teams (CPTs) whose role involves protection of the DoDIN. Each service organizes, trains, and equips cyberspace forces for presentation to CYBERCOM through Joint Force Headquarters (Cyber) (JFHQ[C]) in line with the CMF construct.

Acquisition of Cyber Capabilities

Supporting cyberspace forces requires a steady flow of new technologies and conceptual breakthroughs that ensure forces have the capabilities required to carry out their missions given an ever-changing operational environment. Ideally, the primary source of capability to CYBERCOM and cyberspace forces is the DoD acquisition system.

A complicating factor to cyber acquisition is the distributed and dual-use nature of cyberspace. Shared ownership of this common infrastructure creates a variety of challenges, such as a tension between open engagement and secrecy of cyber operations (related to the need for openness to drive innovation), whether and how to integrate efforts with a private sector that owns and operates large swaths of internet and critical infrastructure, and the proper distribution of mission areas across the broader government cyber enterprise.[2] While the services and CYBERCOM often look to industry for novel concepts, the mixture of Internet Protocol (IP) and

---

1 Ariel Michelman-Ribeiro, "A Primer on the Cyber Mission Force," Center for Naval Analysis (CNA), DIM-2022-U-033387-Final, September 2022. See also Office of the Under Secretary of Defense (Comptroller)/Chief Financial Officer, *Defense Budget Overview: United States Department of Defense Fiscal Year 2023 Budget Request*, April 2022, pp. 2–13; James Di Pane, "Cyber Warfare and U.S. Cyber Command," *2023 Index of Military Strength*, Heritage Foundation, October 18, 2022; Michael Warner, "U.S. Cyber Command's First Decade," Hoover Institution National Security, Technology, and Law Aegis Paper No. 2008, December 3, 2020; and Fiscal Year 2021 Budget Request for U.S. Cyber Command and Operations in Cyberspace, Hearing Before the House Committee on Armed Services, Subcommittee on Intelligence, Emerging Threats, and Capabilities, 116th Congress, March 4, 2020, Statement of Kenneth Rapuano, Assistant Secretary of Defense for Homeland Defense and Global Security and Principal Cyber Advisor.

2 Gen. Paul Nakasone, "CYBERCOM and NSA Chief: Cybersecurity Is a Team Sport," *Defense News*, December 6, 2021.

non-IP-based networks—sometimes referred to as information technology and operational technology, respectively—under CYBERCOM's purview challenges the force to find new and effective ways to innovate, secure, and operate technologies in areas that are traditionally underserved by commercial interests. As a result, current acquisition policy is widely seen as unfit to provide the required support to cyberspace forces.[3]

The strengths of the current system include the diversity and federated nature of cyber capability development to support both service and CCMD needs. This is itself not without challenges, as this decentralized ecosystem can result in a lack of coordination that produces duplication and competing technologies. As a result, DoD has established the Joint Cyber Warfighting Architecture (JCWA) as an organizing paradigm, intended to "integrate systems that enable the cyber warfighting mission."[4] With each service responsible for specific elements of JCWA's implementation, the construct is poised to promote improvements to the cyber acquisition process by more clearly defining development and capability constructs as well as changing the nature of how cyber is employed.

Employment of Cyberspace Forces

The employment of cyberspace forces differs from both the traditional notions of force employment and other cyber conceptions of activity (in structure, if not in effect). Figure 3.1 from *Joint Publication 3-12, Cyberspace Operations* (JP 3-12) depicts the alignment of missions, actions, and forces toward the goal of mission accomplishment.

---

[3] As evidenced by the mandated review of the Federal Acquisition Regulation and defense supplement, as well as associated provisions, contained in Executive Order No. 14028, "Improving the Nation's Cybersecurity," May 12, 2021.

[4] Note that the JCWA is not currently a program but a management construct to oversee the development of distinct but related capabilities. See U.S. Government Accountability Office, "Defense Acquisitions: Cyber Command Needs to Develop Metrics to Assess Warfighting Capabilities," March 30, 2022. However, the National Defense Authorization Act (NDAA) directs the department to establish a JCWA program executive office within CYBERCOM. See National Defense Authorization Act for Fiscal Year 2023, H.R. 7900, 117th Congress (2021–2022), *Congress.gov*, October 11, 2022.

FIGURE 3.1
Cyberspace Operations Missions, Actions, and Forces

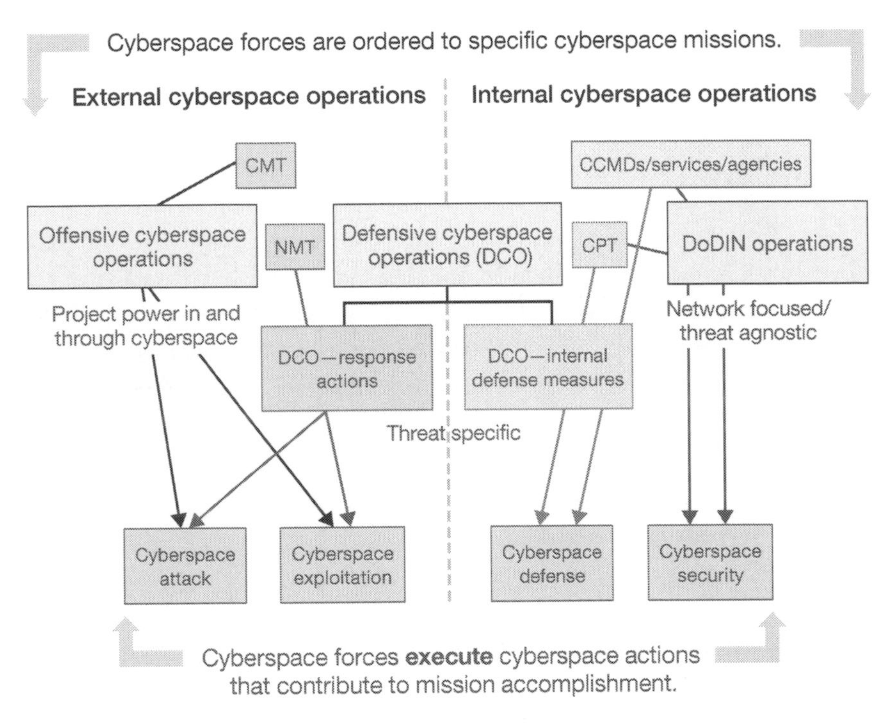

SOURCE: U.S. Department of Defense, *Joint Publication 3-12, Cyberspace Operations*, June 8, 2018, p. II-3.

Inherent in this design is a need to coordinate multiple entities in cyberspace operations. This "professionalization" of cyber warfighting results in structures that minimize the variability of operations and the discretion afforded to operators, providing civilian and military leadership with the structure and processes needed for effective oversight and command.

- **Assessment.** Significant effort is undertaken by CMTs to measure, understand, and express the effectiveness of cyber mission capacity and capability. Assessment ranges from evaluating individual cyber capabilities through operational test and evaluation—for example, Cyber Operations Lethality and Effectiveness (COLE) to support Joint Munitions Effectiveness Manual (JMEM)-like calculations of cyber effects[5]—to certification that each designated individual and team can perform in the warfighting role to which they are assigned. As JCWA develops, this capability will be increasingly supported through the Persistent Cyber Training Environment (PCTE), a flexible and capable cyber environment that supports training, certification, and mission rehearsal.[6]

- **Infrastructure.** To conceal the source and attribution of a cyber effect, skilled actors require a ready supply of resources beyond the planned effect. These resources include
  - *intermediate nodes*, or computers that the attacker controls either by proxy, prior attack, or covert acquisition
  - *accesses*, or means for delivering the effect, which may be as simple as sending an email or more complex, requiring the cultivation of vulnerabilities or other forms of access
  - *personas*, as many effects (especially those supporting information-based operations) require believable online profiles to enable the access or effect
  - *data storage and management*, because cyber missions often involve the generation and acquisition of large amounts of data (offensive or defensive in nature) that must be efficiently stored, searched, and analyzed to support on-mission decisionmaking and post-mission understanding.

  These resources potentially represent a significant investment beyond the cyber effect itself, and for government-sanctioned military operations, development or procurement of these resources may be required under strict conditions.[7]
- **Authorities.** A unique feature of U.S. cyber operations is the distinction between operations conducted for warfighting purposes as opposed to those conducted for other means (such as law enforcement). This

[5] Operational Test and Evaluation, Defense, "Exhibit R-2, RDT&E Budget Item Justification: PB 2021 Operational Test and Evaluation, Defense," February 2020.

[6] PEO STRI, "Persistent Cyber Training Environment (PCTE)," website, undated.

[7] U.S. Government Accountability Office, "Defense Acquisitions: Joint Cyber Warfighting Architecture Would Benefit from Defined Goals and Governance," GAO-21-68, November 19, 2020.

> distinction is rooted in the U.S. system of checks and balances, under which oversight is afforded to operational units based on the title in U.S. law under which the mission is authorized, with the most common being military authority (Title 10) and intelligence authority (Title 50). Additional titles may come into play based on the employment of National Guard units (Title 32) or the need to integrate with law enforcement activities.[8]

These authorities require an extensive level of planning and specification to meet legal requirements. The United States' approach to most cyber operations requires their execution to be covert and leaves the distinction between intelligence gathering and national defense ill defined. These concepts are further clouded by the dual-hat arrangement that has a single commander overseeing both NSA (an intelligence agency) and CYBERCOM (a combatant command), as well as the fluid nature of cyber technologies that allows knowledge and capability to be easily shared and repurposed to meet various needs. As a result, the same personnel, capabilities, and infrastructure may be involved in operations under varying authorities, creating challenges for transparency and oversight.[9] Balancing these authorities often requires extensive legal and policy knowledge in addition to deep technical expertise.

Combined, these elements point to an important aspect of U.S. CYBERCOM operations, which is the rigorous and often time-consuming process by which cyber operations are planned, approved, and executed. Cyber operations should integrate into broader military operations at all levels, following the joint planning process to develop an operational approach that accomplishes mission objectives.[10] *Joint Publication 3-12* additionally identifies several considerations, such as the timeline for execution and unique

---

[8] See U.S. Department of Defense, *Joint Publication 3-12, Cyberspace Operations*, June 8, 2018, p. III-3, for a summary of different U.S. Codes and their role in cyberspace operations.

[9] Andru E. Wall, "Demystifying the Title 10-Title 50 Debate: Distinguishing Military Operations, Intelligence Activities and Covert Action," *Harvard National Security Journal*, Vol. 3, No. 1, 2011, pp. 85–142.

[10] U.S. Department of Defense, *Joint Publication 5-0, Joint Planning*, December 1, 2020.

nature of cyber effects, as well as supporting intelligence and analytic functions required, painting a picture of an integrated cyber force.[11] For offensive operations, targeting follows the same joint targeting cycle[12] employed for all military operations, which flows through commander objectives, guidance, and intent (phase 1), target development and prioritization (phase 2), capabilities analysis (phase 3), commander's decision and force assignment (phase 4), and mission planning and force execution (phase 5). Each step involves the production and consumption of various planning documents; the key documents are those resulting from the cyber force mission alignment process that collects requests from across the CCMDs, turning them into orders that drive the targeting process.

One factor that continues to challenge the full integration of cyber operations is a lack of expertise at the CCMD and component levels.[13] This is a recurring theme, despite the establishment of joint cyber centers (JCCs) and cyber operations–integrated planning elements (CO-IPEs) to standardize command, control, planning, and integration of cyberspace capabilities.[14] As noted previously, factors such as the disparity in the timelines required for the development and employment of cyber capabilities, coupled with the elevated and compartmentalized classifications at which capabilities are often developed, limits the sharing of tools, techniques, and procedures. Recruiting, training, education, retraining, and tracking cyber talent remains an ongoing challenge across the DoD cyber enterprise,[15] leading to

[11] DoD, June 8, 2018.

[12] U.S. Department of Defense, *Joint Publication 3-60, Joint Targeting*, January 31, 2013.

[13] See, for example, Brett Wessley, "Evolution of U.S. Cyber Operations and Information Warfare," *RealClear Defense*, June 10, 2017.

[14] U.S. Government Accountability Office, "Personnel and Support Needed for Joint Cyber Centers," DODIG-2015-048, December 8, 2014.

[15] U.S. Government Accountability Office, "Military Cyber Personnel: Opportunities Exist to Improve Service Obligation Guidance and Data Tracking," GAO-23-105423, December 21, 2022. For the state of cyberspace education, see Quentin E. Hodgson, Charles A. Goldman, Jim Mignano, and Karishma R. Mehta, *Educating for Evolving Operational Domains: Cyber and Information Education in the Department of Defense and the Role of the College of Information and Cyberspace*, Santa Monica, Calif.: RAND Corporation, RR-A1548-1, 2022.

a lack of capable personnel at each echelon and necessarily limiting the level of integration possible.

Even where capability and expertise align, the changing nature of cyber and other nonkinetic operations versus traditional, kinetic operations well known to military commanders results in skeptical views of cyber employment. This may be especially true following the onset of hostilities, as lessons from the Russia-Ukraine conflict point to the changing nature of cyber capability once initial, long-lead, and more exquisite on-network capabilities have been utilized.[16] Advanced capabilities that seek "converged effects" through the integration of cyber with electronic warfare (EW) and other nonkinetic capabilities.[17] However, such capabilities are not yet pervasive across the force, with most operations remaining on-network. This limits the overall scale, scope, and effectiveness that can be achieved in many scenarios and necessitates an evolution in thinking about the nature and role of overmatch in cyberspace.

---

[16] Gavin Wilde and Jon Bateman, "Russia's Wartime Cyber Operations in Ukraine: Military Outcomes and Drivers," CYBERWARCON Conference, November 10, 2022.

[17] Mark Pomerleau, "U.S. Military to Blend Electronic Warfare with Cyber Capabilities," *C4ISRNet*, April 14, 2021.

CHAPTER FOUR

Creating Selective Overmatch in Cyberspace

In this chapter, we present our framework for applying the concept of selective overmatch to the U.S.-PRC competition. The intent of selective overmatch is to allow the United States to defend its national interests while holding Chinese interests at risk with cyberspace capabilities.

Influence Points: Key Elements for National Strength and Survival

Our analysis centers on the concept of influence points. The first step in creating an advantage is to develop a rigorous approach to identifying these influence points. We analyzed the factors or conditions that were essential for the CCP to maintain control of China and for China to prevail in conflict. What we term "policy-level influence points" are elements that are indispensable to a nation's survival. They are supported by political, military, diplomatic, and economic "strategic influence points"; and by critical military, intelligence, and security capabilities that are termed "operational influence points." Operational-level influence points, in turn, are supported by a constellation of links, nodes, facilities, and functions at the tactical level. The United States also has its own policy, strategic, and operational influence points that must be protected in competition and conflict.[1]

[1] The reader may ask how these influence points differ from centers of gravity, which the DoD defines as "the source of power that provides moral or physical strength, freedom of action, or will to act." As Antulio Echevarria has argued, the use of the term

Developing a Framework for Analyzing Influence Points and Actions to Affect Them

The two key variables in understanding influence points are competition versus conflict (peacetime versus wartime operations) and *minimalist* versus *maximalist* intensity. The first variable is well established in joint doctrine.[2] The second captures a range of options for policymakers, strategic decisionmakers, and operational commanders. The minimalist end of the spectrum captures options that are optimized to be measured—largely defensive measures that are less provocative and less escalatory—and actions that are intended to provide effects against the largest number of operational requirements with the minimum number of forces. The "maximalist" end, on the other hand, provides complementary options—more active, offensive, and potentially more provocative and escalatory, requiring a greater military and political commitment to deliver effects against higher-priority influence points.

We captured the competition–conflict spectrum as the x-axis of a graph and the minimalist–maximalist spectrum as the y-axis. The resulting construct, which we refer to as the MC4 chart (for the four types of competition and conflict), allows us to frame strategic and operational choices with respect to the two most relevant variables. Figure 4.1 illustrates this

"center of gravity" has varied across the services, and attempts to reconcile these uses have had mixed results. We use influence points as a term to make clear that it is a vector to exert influence over an adversary, but action against that influence point may not be decisive in itself. Rather, the constellation of influence points is important, whereas center of gravity can imply a single or small group of targets. See Antulio Echevarria, "Clausewitz's Center of Gravity: It's Not What We Thought," *Naval War College Review*, Vol. 56, No. 1, Winter 2003. For the DoD definition, see U.S. Department of Defense, *Dictionary of Military and Associated Terms*, Washington, D.C.: The Joint Staff, November 2021.

[2] *Joint Doctrine Note 1-19, Competition Continuum*, June 3, 2019, p. v, states that "rather than a world either at peace or at war, the competition continuum describes a world of enduring competition conducted through a mixture of cooperation, competition below armed conflict, and armed conflict." While the formal intermediate step of "crisis" is used less often, joint doctrine now employs "competition" and "conflict" as anchoring terms along the spectrum of conflict.

FIGURE 4.1
The MC4 Construct: Postures in Competition and Conflict

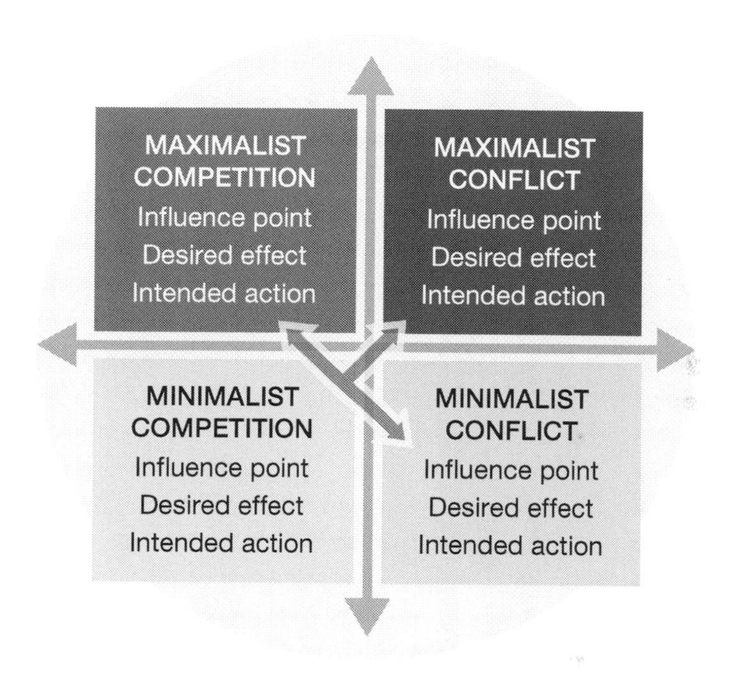

alignment of minimalist and maximalist spectrums in competition and conflict.

Influence points do not reside exclusively in a quadrant; rather, most influence points can be affected by actions in each of the quadrants. To develop targets in each quadrant, we determined it was necessary to ask, "What is the desired effect against this influence point in this quadrant?" and second, "What intended actions are required to achieve each desired effect?" For example, if the policy-level influence point is the CCP regime's survival, then a desired effect in the minimalist conflict quadrant may be to reduce the regime's ability to track people internally, and an intended action to achieve this effect could be to degrade the functioning of the Great Firewall and the stability maintenance system. In Chapter 5, we discuss the process of identifying and prosecuting influence points in greater detail and provide examples of the process.

Creating Capabilities to Affect Influence Points: Evolving Cyberspace Operations

The next step in developing our operational concept was to examine current CYBERCOM capabilities against this hierarchy of intended actions to produce desired effects against specific influence points within the policy, strategic, and operational limitations of each quadrant of the MC4 chart. We conducted this analysis at the unclassified level, examining media, academic, and DoD public affairs sources to evaluate categories or group of capabilities, as opposed to looking at classified single instances involving specific tools employed with specific accesses. We believe this gave us an academically rigorous, operationally relevant constellation of capabilities to evaluate in the context of the operational concept we were developing.

We identified capability gaps against the large set of targets we evaluated. These gaps fall into four broad categories for future expansion and improvement. We designated these areas as integrated cyber, integral cyber, expanded cyber, and extended cyber, which we refer to collectively as the I2E2 concept.

Integrated cyber is a "down and in" concept that incorporates adjacent capabilities (cyber-cognitive in addition to cyber-technical, organic intelligence more enduring than the NSA-CYBERCOM "dual hat" and certain information-related and EW capabilities) to provide all the tools required by the operational cyber commander. Although this concept does not call for the complete doctrinal absorption of these capabilities under cyberspace, it provides the full suite of forces needed to produce coherent effects under a unified command. A historical analogy would be the incorporation of organic airpower by Marine Corps expeditionary units. This did not remove the need for tailored Army or Navy aviation, and it certainly did not displace the Air Force as the professional home for large-scale airpower (i.e., strategic airlift, strategic intelligence, surveillance, and reconnaissance (ISR), global precision strike), but it did demonstrate that absolutely siloed warfare areas are not optimal in providing tailored, responsive tactical and operational support in all areas. In addition, integrated cyber allows a single cyber commander to bring weapons from all domains to bear in a single cyber operation or campaign. At present, CYBERCOM has limited organic information and electronic warfare capabilities. In addition, there is a bright policy line between cyber and information within the DoD, though there are approval processes to bring the two together. Cyberspace and other information-related capabilities are under the policy direction of two different assistant secretaries of defense and the operational direction of two different combatant commands (CYBERCOM and U.S. Special Operations Command).

Integral cyber is an "up and out" concept that recognizes that any significant military operation in the twenty-first century is unthinkable without necessary cyber components (i.e., encrypted communication; position, navigation, and timing; sensor integration; multiplatform targeting) that greatly enhance speed, precision, and lethality. These same cyber capabilities may be exploited by a sophisticated adversary, leaving a modern force severely limited and vulnerable to neutralization. As China continues to modernize and expand its multidomain forces, it becomes both more capable and simultaneously more vulnerable. As new weapons, platforms, and operational concepts are developed, this will become more and more the case. The People's Liberation Army (PLA) will demand cyber as an *integral component* to safeguard and enable its forces. Here, Operation Orchard is a useful historical analogy. In 2007, Israeli air forces struck a suspected Syrian nuclear research and development facility. The attack was made feasible by Israeli cyber forces, which deceived and degraded Syrian air defenses during the raid. Cyberspace was included in the planning, rehearsal, and execution of the mission and provided an early example of effectively integrated cyber and kinetic forces.[3] Currently, CYBERCOM has only limited participation in operational planning with the other combatant commands (primarily through CYBERCOM elements assigned to each combatant command known as cyberspace operations integrated planning element and the JFHQ(C) assigned to support each CCMD) and similarly limited opportunity to train and exercise with cyber capabilities in all-domain operations.

Expanded cyber is built on two key premises. First, cyberspace operations in the future will expand beyond traditional peacetime constraints in the minimalist-competition quadrant to include more assertive action in the maximalist-competition quadrant and must be prepared to fully inhabit

[3] Cooperative Cyber Defense Center of Excellence (CCDCOE), "Operation Orchard/Outside the Box (2007)," *Cyber Law Toolkit*, September 6, 2007.

the minimalist and maximalist quadrants in conflict. Second, and perhaps more important, it will require strategic and operational agility to move from quadrant to quadrant more quickly than the adversary. This will require the situational awareness that emerges from persistent engagement and the operational flexibility that comes from continuous planning, training, and exercising of both integrated cyber capabilities and cyber capabilities integral to all-domain operations. Expanded cyber will allow U.S. cyberspace forces to seize and maintain the initiative with tighter observe-orient-decide-act loops and a broader range of strategic and operational cyberspace options. Further research in this area may yield more insights into the value of producing larger numbers of cyberspace capabilities and accesses that are broadly applicable against targets to complement the exquisite weapons and accesses that predominate today. At present, CYBERCOM's doctrine of persistent engagement is fully consonant with this vision of more agile operations, but current force structure requires strict prioritization against most likely, immediate, and potentially damaging threats.

Extended cyber broadens CYBERCOM's operational horizons across another dimension and beyond the active technical networks that have been the focus of operations up to this point. Extended cyber now includes visibility across supply chains, life cycles, and emergent networks that were not designed as a coherent whole. This provides both a broader view of threats and more opportunities for neutralizing them. Today, China and other adversaries are exploiting and attacking cyber targets from the time of their design through their eventual retirement.[4] They are approaching them along numerous vectors, only some of which are currently within CYBERCOM's authority to observe and interdict. Current roles, mission, authorities, and capabilities across the federal government are too fragmented to provide comprehensive situational awareness or coherent action over time or across vulnerable networks of networks, such as the defense industrial base, supporting critical infrastructure, and cognitive rather than technical cyber targets.

[4] For example, Chinese cyber-enabled espionage appears to have aided its economic development and the acquisition of advanced military technologies. See Office of the National Counterintelligence Executive, *Foreign Spies Stealing U.S. Economic Secrets in Cyberspace: Report to Congress on Foreign Economic Collection and Industrial Espionage, 2009–2011*, October 2011. See also Office of the Director of National Intelligence, *Annual Threat Assessment of the U.S. Intelligence Community*, February 6, 2023.

CYBERCOM may not be the supported command in addressing all these threats, but an effective allocation of responsibility from the national to the tactical level requires a more seamless understanding of the U.S. attack surface vulnerable to China. Currently, this is the most challenging area for CYBERCOM because of the complexity of the larger U.S. cyber ecosystem and numerous statutory and policy limitations on CYBERCOM playing a larger domestic role. Support to defend elections since 2018 has provided useful lessons for effective ways ahead in this area.[5]

Conducting Cyber Operations Using Selective Overmatch

To use selective overmatch as an operational concept, it is first necessary to identify adversary influence points at the policy, strategic, and operational levels to be targeted in competition and conflict within the context of United States national security objectives and subordinate guidance (we discuss specific Chinese and U.S. influence points in Chapters 5 and 6, respectively). Next, it calls for a detailed mapping of desired effects against each influence point and then mapping intended actions to achieve each desired effect.

This mapping is done through the four perspectives of minimalist competition, maximalist competition, minimalist conflict, and maximalist conflict. Each posture frames a different set of tools, actions, and restraints against any particular influence point. For example, reconnaissance may be undertaken against an influence point in the minimalist-competition phase. Actions in this phase would be designed to be minimally detectable and to be as non-escalatory as possible even if detected. In the maximalist-competition phase, more aggressive preparation of the cyber battlefield may be undertaken, possibly including remote accesses and leave-behind rootkits. These operations may be more detectable and may be more provocative if detected. Such actions undertaken with the intent of being discovered may signal commitment and capability to an adversary.

---

[5] C. Todd Lopez, "Cyber Command Expects Lessons from 2018 Midterms to Apply in 2020," *Defense.gov*, February 14, 2019.

The legal and policy framework shifts dramatically with the movement from competition to conflict. In minimalist conflict, economy of force operations designed to reduce the chance of escalating the conflict degrade one target or produce reversible effects in another. Finally, in maximalist conflict, capabilities are employed to produce irreversible effects, destroying critical links and nodes.

After planning multiple intended actions to produce multiple desired effects against multiple influence points from each of the four MC4 Chart quadrants, it will become clear that present-day CYBERCOM roles, missions, authorities, and capabilities will not support many required actions in and through cyberspace. Selective overmatch highlights four broad areas of growth to permit a single cyber commander to undertake all intended cyber action to produce all desired effects against all influence points.

Integrated cyber will give that commander a more robust and fully formed arsenal, including organic information, EW, and other cyber-adjacent capabilities. Integral cyber will ensure that kinetic and cyber forces train, plan, exercise, and operate together to the greatest possible degree, rather than merely deconflicting operations in separate domains. Expanded cyber will give strategic decisionmakers and operational commanders situational awareness across all four quadrants (minimalist competition, maximalist competition, minimalist conflict, and maximalist conflict), allowing them to maintain the status quo in one quadrant or to seize the initiative by choosing to move to another quadrant. Extended cyber will add the final dimension, giving commanders, planners, and operators visibility and access to systems over their entire life cycle as well as emergent networks on each level of cyberspace—geographic, technical, logical, human, and cognitive.

In the near term, selective overmatch will provide a logical, rigorous, and transparent framework for more effective force employment. In the longer term, it will provide a clear, principled, and objective basis for deliberate force design. Selective overmatch complements the doctrine of persistent engagement, providing tools to determine the type and degree of engagement required to meet strategic and operational objectives. It is a means by which military requirements may be communicated to non-cyber-expert leadership on the Joint Staff, in the Office of the Secretary of Defense, the White House, and Congress, and through the press to the American people.

Selective overmatch provides a basis for "intellectual interoperability" with allies and partners, allowing countries with differing force structures, doctrine, and levels of capability to cooperate in planning and executing cyber operations. By integrating them more effectively in multidomain operations and widening the aperture of cyber operations, selective overmatch permits a more balanced approach to risk management from acquisition to personnel assignments. Finally, it will heighten the visibility and clarity of cyberspace as a complex, adaptive system-of-systems, allowing planners and advisers to better distinguish complexity and chaos.

Selective overmatch will drive change for CYBERCOM, some of it foreseeable but much of it not. While persistent engagement may remain the firm doctrinal foundation for cyber operations, many other variables, from the number and composition of mission teams to the roles and missions and authorities given to the commander, may be dramatically altered. It will change not just cyber operations but the kinetic operations that depend on cyber capabilities. Selective overmatch, properly understood and applied, can provide a roadmap to CYBERCOM's future, illuminating the way ahead.

CHAPTER FIVE

Identifying Chinese Influence Points and Actions to Affect Them

This chapter identifies specific influence points relevant to the PRC and examines one of those influence points in greater detail under different states of competition and conflict. The specific influence point examined is the CCP's information control system-of-systems, frequently referred to as the Great Firewall (GFW) by those observing from outside the PRC. This influence point is uniquely important because it a cornerstone of current stability management practice within the PRC. After reviewing the GFW and potential U.S. actions under various states of competition and conflict, we then conclude with a brief discussion of how to continue this research, and important caveats about what this exploratory research has presented.

The PRC has numerous key influence points at the policy, strategic, and operational levels. Disrupting any influence points is likely to sow doubts within CCP leadership circles about their ability to maintain control over Chinese citizens and the PLA—both of which are critical, from Beijing's perspective, to ensuring regime security.

At the policy level, China prioritizes the following objectives in the following order: (1) maintain exclusive CCP control; (2) ensure domestic social stability; (3) continue overall economic development; (4) advance domestic industry, technology, and science; (5) establish a global mercantile network for secure access to foreign critical materials; (6) establish Sino-centric order in the Indo-Pacific region; and (7) develop a military capable of defending all the preceding goals. We address the nature of each of these policy-level influence points in the next sections.

Exclusive CCP Political Control

On the first policy influence point, the CCP in 1949, under the first paramount leader Mao Zedong, established the PRC, and since then, it has refused to allow any viable political opposition to exist in the system.[1] With more than 90 million members, the CCP has penetrated Chinese society broadly and deeply, contributing to the party's durability.[2] Since Chinese President Xi Jinping came to power in 2012, the CCP has only become more repressive, whether against ethnic Uyghurs in Xinjiang, Chinese business leaders, Chinese citizens residing in the PRC and living abroad, or even Xi's own colleagues in power.[3] At the CCP's 20th Party Congress, held in October 2022, Xi received an unprecedented third term in office. This, coupled with the fact that he no longer faces term limits on his tenure, suggests that Xi could serve as leader for life.

To ensure domestic social stability, the CCP has invested heavily in building a surveillance state to monitor its citizens and maintains police, paramilitary, and military forces that might be needed for domestic contingencies. CCP investments in societal control, however, have not prevented unrest. In November 2022, for example, a significant backlash against the CCP's strictly enforced "Zero COVID" policy resulted in protests across the PRC. According to leaked CCP directives, Chinese authorities initiated the highest "emergency response" level of censorship to prevent the protests

---

1 To be sure, there are eight legally permitted political parties outside of the CCP. However, all are controlled by the CCP's United Front, and they are only allowed to offer legislative proposals to the CCP through the Chinese People's Political Consultative Conference (CPPCC)—a body with no legislative power. For more, see Shannon Tiezzi, "What Is the CPPCC Anyway?" *The Diplomat*, March 4, 2021.

2 Lindsay Maizland and Eleanor Albert, "The Chinese Communist Party," Council on Foreign Relations, October 6, 2022.

3 Lindsay Maizland, "China's Repression of Uyghurs in Xinjiang," Council on Foreign Relations, September 22, 2022; Kana Inagaki, Leo Lewis, Ryan McMorrow, and Tom Mitchell, "Alibaba Founder Jack Ma Living in Tokyo Since China's Tech Crackdown," *Financial Times*, November 29, 2022; "China Builds a Self-Repressing Society," *The Economist*, May 14, 2022; "Special Report 2021: China: Transnational Repression Origin Country Case Study," Freedom House, undated; Rahul Karan Reddy, "China's Anti-Corruption Campaign: Tigers, Flies, and Everything in Between," *The Diplomat*, May 12, 2022.

from gaining steam. In particular, the CCP instructed operators of its Great Firewall to crack down on virtual private networks (VPNs) and other mechanisms of avoiding censorship.[4] The CCP had additional concerns over the coincidental timing of former leader Jiang Zemin's death and how protesters might seek to juxtapose his professional record, which is widely regarded among young Chinese as relatively benign and good for China, with Xi's perceived incompetent, reckless, and dictatorial style. As an illustration, just days before Xi received his third term at the 20th Party Congress, a man draped a banner over a busy intersection that called for students and workers to strike and remove the "dictator and state traitor Xi Jinping," inspiring others to protest as well.[5] These incidents appear to have been handled swiftly, if not ruthlessly, by Chinese authorities. Indeed, the CCP has deep experience and, probably, confidence, in defusing tense situations as the nation reportedly deals with tens of thousands of protests per year.[6]

Social Stability Rooted in Economic Development

For CCP leaders, continued economic development is the driver of social stability and ultimately the guarantor of the regime. If the people are unhappy, then the CCP typically attributes it to deficient "socioeconomic development." A good example was the CCP's response to societal unrest in 2019 in Hong Kong. In addition to strengthening its security presence on the ground in Hong Kong through the National Security Law passed in 2020, Xi's speech at the 20th Party Congress emphasized the importance of growing Hong Kong's economy and improving the living standards there as part of maintaining stability.[7] China has embarked on integrating Hong

[4] Helen Davidson, "China Brings in 'Emergency' Level Censorship over Zero Covid Protests," *The Guardian*, December 2, 2022.

[5] Yibing Feng, "Beijing Banner Protest Ripples Outward as China Maintains Silence," *Voice of America*, October 20, 2022.

[6] Teresa Wright, "Protests in China Are Not Rare—but the Current Unrest Is Significant," *The Conversation*, November 30, 2022.

[7] Ministry of Foreign Affairs, "Full Text of the Report to the 20th National Congress of the Communist Party of China," October 25, 2022, pp. 50–51.

Kong into its Greater Bay Area scheme that will include Guangdong and Macau, in the hopes that Hong Kong reemerges as a global hub of commerce. Ultimately, the goal is to satiate Hong Kongers by elevating their livelihoods rather than upholding and strengthening their political rights. The CCP uses this same model for mainland China. However, mass shutdowns caused by the Zero COVID policy resulted in slumping economic growth. In 2022, the World Bank assessed China's real GDP growth to be around 3 percent for the first nine months and approximately 2 percent for the last three months—well below the official target of 5.5 percent.[8] One reason for the CCP's sudden reversal of Zero COVID in December 2022 and the full reopening of cities was probably concerns over lackluster economic growth and the potential impact on social stability and regime control.[9]

Continuing Scientific Innovation

Related to economic growth, the CCP seeks to advance domestic industry, technology, and science. During his speech to the 20th Party Congress, Xi was clear that his China will continue to rely on innovation—not only to become a scientific and technological great power alongside the United States and Russia, but also to create new jobs as the PRC increasingly becomes a modernized economy.[10] Although the PRC can adopt Western concepts and technologies by adding "Chinese characteristics" to them, Xi's goal, as outlined in the speech, is to chart a path of "self-reliance." This is quite significant because it indicates that Beijing's primary focus will be indigenous innovation to avoid scientific and technological reliance on outside powers that may seek to undermine or prevent China's growth.

---

[8] "China GDP Annual Growth Rate," *Trading Economics*, undated.

[9] "China's Xi Bets Economic Growth Will Offset Misery from Covid," *Bloomberg News*, December 22, 2022.

[10] Ministry of Foreign Affairs, October 25, 2022, p. 28.

A Global Mercantile Network

The CCP further seeks to establish a global mercantile network for secure access to foreign critical materials. In simple terms, Beijing is prioritizing the import of energy and natural resources (primarily though not exclusively from Africa and the Middle East) necessary to fuel China's economic growth. Beijing requires a network of access points, both on land and at sea, to facilitate these activities. One proven method is to leverage China's Belt and Road Initiative (BRI). BRI is a global investment and infrastructure program that is attractive to developing world nations. Beijing prioritizes BRI engagement with those nations that possess, for example, rich oil or natural gas deposits.[11] China also prioritizes those with vast mineral or fishery reserves, among many other critical materials. BRI provides an entree into developing countries that in several cases has resulted in port access and other preferential agreements.

Establishing a Sino-Centric Indo-Pacific Order

From a broader perspective, the CCP wishes to establish—or, perhaps more accurately, reestablish—a Sino-centric order in the Indo-Pacific region. According to Xi's "China Dream," the PRC can reverse the Century of National Humiliation—stretching from when China was semicolonized starting in the late 1830s until the establishment of the PRC in 1949—by actively pursuing the Century of National Rejuvenation out to 2049.[12] According to this narrative, China will once again become powerful, as it was during the Ming and Qing dynasties. During the Ming and Qing dynasties, considered the golden periods of Chinese history, China was at the center of Asia, with smaller states such as Korea, Japan, and Vietnam engaged in the Chinese tributary system. By 2049, Xi and the CCP probably do not expect a resurrection of the tributary system, but they certainly seek to reposition China as the dominant force on the Asia continent. Indeed,

---

[11] Muhammad Zulfikar Rakhmat, "The Belt and Road Initiative in the Gulf: Building 'Oil Roads' to Prosperity," Middle East Institute, March 12, 2019.

[12] "What Does Xi Jinping's China Dream Mean?" *BBC News*, June 6, 2013.

Beijing for decades has been frustrated by the fact that former tributary states Korea (South Korea) and Japan have security treaties with the United States that, from the CCP's perspective, are being leveraged to "contain" China's growing power.[13] In an ideal geostrategic landscape for Beijing, neither the United States nor any other outside power would rival China within the Indo-Pacific.

Developing a Military Capable of Promoting Economic and Political Goals

Finally, the CCP seeks to develop a military capable of defending all the preceding goals. The PRC's premier security force is the PLA; and it is comprised of five services including the PLA Army (PLAA), PLA Navy (PLAN), PLA Air Force (PLAAF), PLA Strategic Support Force (PLASSF), and PLA Rocket Force (PLARF). Beijing for decades has prioritized the modernization and professionalization of the PLAA, PLAN, and PLAAF, particularly, though not exclusively, to conduct a successful amphibious-landing invasion of Taiwan. The CCP probably believes that maintaining regime security to some extent relates to having the capabilities to conquer Taiwan, which Beijing considers an immutable part of China, if that action is ever deemed necessary.

In late 2015, Xi announced PLA reforms that not only changed the structure and way in which the PLAA, PLAN, and PLAAF function but also added the PLASSF and elevated and renamed the Second Artillery to the PLARF.[14] The PLASSF's role is to support all other services through the integration of cyberspace, space, and electromagnetic data, while the PLARF manages China's ground-launched missile systems, including nuclear-armed inter-

[13] China has used the term "containment" for many years to describe U.S. policy. See, for example, Wang Yi, Ministry of Foreign Affairs, "The Right Way for China and the United States to Get Along in the New Era," speech at the Asia Society, New York, September 22, 2022.

[14] Andrei A. Kokoshin, *2015 Military Reform in the People's Republic of China: Defense, Foreign, and Domestic Policy Issues*, Belfer Center Paper, Belfer Center for Science and International Affairs, October 2016.

continental ballistic missile (ICBMs).[15] Beijing also has two other PLA components at sea: the Chinese Coast Guard (CCG) and People's Armed Forces Maritime Militia (PAFMM). The CCG patrols Beijing's expansive "nine-dash line" claims in the South China Sea, which account for approximately 90 percent of this body of water. Many of these regions are disputed, as they overlap with counterclaimants' exclusive economic zones, which include Brunei, Malaysia, the Philippines, Taiwan, and Vietnam. Although not an official counterclaimant, Indonesia has also expressed concerns about the implications of Chinese claims for its security. Hence, the CCG is hardly a purely constabulary force. It reports directly to the PLA and has the authority, as of January 2021, to fire on rival vessels if necessary.[16] The PAFMM are fishermen who are trained to use light arms and to conduct surveillance operations on behalf of the PLA. They have played an integral role in China's establishment of a de facto operating presence in many disputed regions of the South China Sea and elsewhere.[17] In addition to the PLA, China also maintains the People's Armed Police (PAP) to handle terrorist and social unrest incidents. The PAP bolsters local law enforcement operations. And finally, the Ministry of State Security (MSS) and Ministry of Public Security (MPS) are actively involved in spying on and detaining any potential threats to the regime. The MSS handles threats emanating from abroad, while the MPS focused its efforts domestically.

China's Strategic Influence Points

At the strategic level, China prioritizes the following objectives as strategic influence points, in rank order: (1) preserve the reputation and legitimacy

[15] For further analysis on the PLASSF, see Kevin L. Pollpeter, Michael S. Chase, and Eric Heginbotham, *The Creation of the PLA Strategic Support Force and Its Implications for Chinese Space Operations*, Santa Monica, Calif.: RAND Corporation, RR2058-AF, 2017. For further analysis on the PLARF, see Ma Xiu, "PLA Rocket Force Organization: Executive Summary," China Aerospace Studies Institute, November 29, 2021.

[16] Shigeki Sakamoto, "China's New Coast Guard Law and Implications for Maritime Security in the East and South China Seas," *Lawfare*, February 16, 2021.

[17] Derek Grossman and Logan Ma, "A Short History of China's Fishing Militia and What It May Tell Us," The RAND Blog, April 6, 2020.

of the CCP leadership; (2) ensure efficacy of Politburo Standing Committee (PBSC) and Central Military Commission (CMC); (3) maintain integrity of PLA strategic forces; (4) gain conventional military primacy within China's historical zone of hegemony; (5) strengthen BRI to ensure continuing access to critical materials; (6) develop PLA expeditionary forces to defend regional hegemony and global commercial interests; (7) preserve strategic situational awareness and decisionmaking ability through C5ISRT, artificial intelligence (AI), and decision support technology; and (8) maintain capability to influence and degrade adversary strategic decisionmaking ability. These strategic objectives are derived from a review of Chinese official statements and documents, and this description builds on prior RAND research on Chinese strategic thinking.[18]

It is of paramount importance for the CCP to preserve its reputation and legitimacy within the system. CCP leaders believe that regime security could be jeopardized if its reputation and legitimacy are called into question. A good example of an incident that challenged perceptions of the CCP was the release of the so-called Panama Papers. The Panama Papers were a collection of 11 million leaked documents from a Panamanian law firm that were subsequently analyzed by the International Consortium of Investigative Journalists and reported on by media outlets around the world. Regarding the CCP, documents in the collection indicated that the family members of at least eight current and former PBSC members, including Xi, maintained secret offshore companies.[19] Although these allegations were hardly the bombshells needed to shake CCP confidence in its ability to control the outcome, the Panama Papers offered an insider glimpse into the level of corruption going on within the party's ranks. The CCP was able to handle the leak effectively through online censorship.[20] However, future leaks could

[18] See, for example, Andrew Scobell, Edmund J. Burke, Cortez A. Cooper III, Sale Lilly, Chad J. R. Ohlandt, Eric Warner, and J. D. Williams, *China's Grand Strategy: Trends, Trajectories, and Long-Term Competition*, Santa Monica, Calif.: RAND Corporation, RR-2798-A, 2020.

[19] Andrew Nathan, Bill Bishop, David Wertime, and Taisu Zhang, "China in the Panama Papers," ChinaFile Conversation, April 6, 2016.

[20] Tom Phillips, "All Mention of Panama Papers Banned from Chinese Websites," *The Guardian*, April 5, 2016.

reveal more explosive and less controllable derogatory information about the party.

The CCP also endeavors to ensure the efficacy of the PBSC and CMC. Without an appropriate level of coordination and communication among and between these bodies, the prospect for strategic mistakes rises. In the wake of the 20th Party Congress, one of the concerns is that Xi stacked the PBSC with loyalists who probably cannot reach independent conclusions on policy or conclusions that would run counter to Xi's own approach.[21] Having attained the status of "core leader" within the CCP, Xi is now the unquestionable authoritarian who is increasingly ruling the PRC in a personalized and totalitarian fashion. Xi has successfully achieved an unprecedented third term, got the party to eliminate term limits, and enshrined his ideology, known as "Xi Jinping Thought," alongside the last PRC strongman and founder Mao (Mao Zedong Thought). Before Xi's ascension, and especially in the Jiang Zemin and Hu Jintao periods, collective decision-making within the PBSC was the common practice. This is not to say that Jiang or Hu did not have an outsized voice or the final say, but at least there was some level of discussion and occasional dissent among PBSC members. Xi has completely gutted today's PBSC of these features. His snap decision to reverse the three-year-old Zero COVID policy and to unleash the chaos that ensued illustrates the dangers of one-man, unquestioned leadership. Xi has similarly packed the CMC with loyalists, making it difficult to consider military decisions that would contradict his preferences.[22]

Separately, the CCP seeks to maintain the integrity of PLA strategic forces. A key component of China's deterrence strategy against the United States is the potential employment of nuclear weapons via ground, sea, and air-based systems. Although Beijing maintains an official "No First Use" policy, its race to build more nuclear weapons—perhaps as many as 1,000 operational nuclear warheads by 2030 (up from the low-200s in 2020)—raises questions about whether such a policy pronouncement can be

[21] "Meet China's New Politburo Standing Committee," Mercator Institute for China Studies, November 17, 2022.

[22] Lyle Morris, "What China's New Central Military Commission Tells Us About Xi's Military Strategy," Asia Society Policy Institute, October 27, 2022.

trusted.[23] China is also pursuing the use of multiple independently targetable reentry vehicles as part of its nuclear warhead production, complicating efforts to counter them.[24] Beijing continues to reject American requests for arms control negotiations.[25] But it is not just the number or even capabilities of China's nuclear warheads. The PLARF is actively modernizing and adding ICBM launchers to its inventory to provide adequate nuclear delivery systems.[26] All of this suggests that CCP leaders view PLA nuclear forces as integral not only to PRC deterrence efforts in a variety of contingencies, most notably a Taiwan scenario, but also as bolstering the PRC's narrative that it indeed is a "great power" that deserves consideration alongside the United States and Russia.

Below the nuclear threshold, the PRC also seeks to gain conventional military primacy within China's historical zone of hegemony. For example, Beijing now numerically fields the world's largest naval force, with approximately 340 ships and submarines. The PRC's aviation and naval aviation forces are the largest in the Indo-Pacific and the third largest in the world, with over 2,800 total aircraft.[27] Meanwhile, the PLARF in 2021 tested 135 conventional missiles, representing more than the rest of the world combined, excluding missiles used in war zones.[28] Beijing's CCG is also the world's largest maritime law enforcement force.[29] In totality, China hopes to deter U.S. military intervention in a variety of scenarios, ranging from the conquest of Taiwan to seizing or retaining disputed features against counterclaimants in the South China Sea or East China Sea and combating India along their disputed border in the Himalayas.

The CCP wishes to strengthen BRI to ensure continuing access to critical materials. As noted above, BRI is an attractive program for developing

---

[23] U.S. Department of Defense, *Military and Security Developments Involving the People's Republic of China*, Annual Report to Congress, November 29, 2022, p. 97.

[24] DoD, November 29, 2022, p. 65.

[25] DoD, November 29, 2022, p. 97.

[26] DoD, November 29, 2022, p. 65.

[27] DoD, November 29, 2022, p. vi.

[28] DoD, November 29, 2022, p. vii.

[29] DoD, November 29, 2022, p. 78.

world countries because it provides investment and infrastructure development assistance. Through BRI, China places a priority on engaging nations that possess rich energy and natural resources reserves.

Beijing believes it must also develop PLA expeditionary forces to defend regional hegemony and global commercial interests. According to authoritative documents, such as the report of the 20th Party Congress or the 2019 white paper, as China increasingly becomes a global power, its interests will naturally expand, necessitating an additional security presence to protect these new overseas interests.[30] China's new security presence typically entails a combination of security personnel, such as defense attachés and intelligence agents, stationed in host countries as well as the deployment of additional military assets, primarily PLAN ships, to assist in securing Chinese interests. Beijing first began expeditionary deployments in the late 2000s when Somali piracy in the Gulf of Aden, which lies between the Suez Canal and the Indian Ocean, affected multiple nations' commercial activities, including those of China. Beijing responded by deploying PLAN to the multinational mission to conduct anti-piracy operations. However, once the threat was defeated and the mission ended, Beijing continued its deployments to the region, culminating in 2017 when it announced the opening of its first official overseas base in the African nation of Djibouti. The PRC claimed that it needed a maintenance and logistics hub in the region to support its operations. In effect, Djibouti became a test case for whether Beijing could operate an overseas base to protect its expanding global interests; the answer appears to be yes, as it is building a new base—or two—in Cambodia and reportedly has interests in other countries such as United Arab Emirates, perhaps the Solomon Islands, and others.[31]

---

[30] Ministry of Foreign Affairs, October 25, 2022, p. 28; State Council Information Office, "China and the World in the New Era," white paper, September 27, 2019. For additional analysis on this topic, see Timothy R. Heath, Derek Grossman, and Asha Clark, *China's Quest for Global Primacy: An Analysis of Chinese International and Defense Strategies to Outcompete the United States*, Santa Monica, Calif.: RAND Corporation, RR-A447-1, 2021, pp. 63–64.

[31] Cristina L. Garafola, Timothy R. Heath, Christian Curriden, Meagan L. Smith, Derek Grossman, Nathan Chandler, and Stephen Watts, *The People's Liberation Army's Search for Overseas Basing and Access*, RAND Corporation, Santa Monica, CA, RR-A1496-2, 2022.

The CCP seeks to preserve strategic situational awareness and decision-making ability through C5ISRT, AI, and decision support technology. In the military realm, Beijing is leveraging the PLASSF to collect, analyze, and disseminate cyber, space, and electromagnetic data in support of the other services. On the domestic side, the CCP is actively investing in building tools, such as supercomputers, to help it make better decisions. Ultimately, the goal is to move from big data analytics, which involves computers analyzing large datasets, to AI, which involves allowing computers to make decisions on behalf of humans. Although the United States and other nations are uncomfortable with this thought, China embraces AI because it fully believes that computerized decisionmaking is inherently more accurate than human decisionmaking. Fewer errors can occur with computers when human emotions and cognitive limitations are eliminated, according to Beijing's perspective.[32]

The last strategic-level influence point for the CCP is to maintain the capability to influence and degrade adversary strategic decisionmaking ability. Referred to in PLA literature as "cognitive domain operations," the PLA plans to employ a holistic approach to psychological operations that essentially controls or conditions an adversary's mindset, making it more malleable or breakable across the spectrum of peacetime, crisis, and wartime scenarios.[33] The PLA's cognitive domain operations also encompass public opinion coercion in peacetime, as was observed during Taiwan's local elections in 2018 and the Hong Kong protests in 2019. During both events, China deployed its narrative through propaganda and other mechanisms to influence the outcomes, with questionable levels of success. Nevertheless, Beijing increasingly believes that "mind superiority" or "brain control," according to one PLA publication, is an attainable and justifiable goal.[34]

---

[32] Derek Grossman, Christian Curriden, Logan Ma, Lindsey Polley, J. D. Williams, and Cortez A. Cooper III, *Chinese Views of Big Data Analytics*, RAND Corporation, Santa Monica, CA, RR-A176-1, 2020.

[33] Nathan Beauchamp-Mustafaga, "Cognitive Domain Operations: The PLA's New Holistic Concept for Influence Operations," *Jamestown Foundation China Brief*, Vol. 19, No. 16, September 6, 2019.

[34] Discussed in Beauchamp-Mustafaga, 2019.

Operational-Level Priorities

Finally, at the operational level, China prioritizes the following objectives, in rank order, as points subject to influence: (1) maintain the Great Firewall and stability maintenance system; (2) defend the strategic mobilization system; (3) preserve PLASSF capabilities; (4) maintain the capability to hold adversary national critical functions and infrastructure at risk; (5) defend theater command headquarters and command posts; (6) preserve deployable major force groups to seize and maintain the military initiative; (7) delay adversary contingency deployment to region; (8) hold adversary forward-deployed forces, facilities, and materiel at risk; and finally, (9) deny regional access to adversary expeditionary forces. These operational-level influence points are derived from the analysis of Chinese doctrine and policy and from prior RAND work addressing Chinese military operations and forces.

Beijing's investments in building and maintaining the Great Firewall is covered in greater detail below as a case study. Suffice it to say here that the firewall is the CCP's shield against online criticism and the public's efforts to mobilize against it, which supports maintaining social stability and the exclusive role of the CCP in governing the country. Additionally, the firewall aims to prevent the Chinese public from having full access to the outside world because the CCP fears the consumption of foreign concepts that could undermine or threaten its survival.[35] The Great Firewall is not impermeable—it is decidedly porous, partly by design.[36] It can be bypassed through creative communication, such as by posting memes or using Chinese characters with dual or hidden meanings, as well as through secure VPNs. In this regard, the GFW is more "friction" than barrier. The CCP supplements this "friction" technical barrier with a combination of "fear" and "flooding": intimidation and propaganda.[37] However, the stability management system is a much broader program encompassing not only the

[35] Stephanie Yang, "As China Shuts Out the World, Internet Access from Abroad Gets Harder Too," *Los Angeles Times*, June 23, 2022.

[36] Margaret E. Roberts, *Censored: Distraction and Diversion Inside China's Great Firewall*. Princeton, N.J.: Princeton University Press, 2018, pp. 224–227.

[37] Roberts, 2018, pp. 80–81.

Great Firewall but many other CCP mechanisms to ensure social and political stability. One such method is by providing adequate employment so that people will not resort to crime or other behavior that could cause trouble. Besides carrots, the stability management system also employs sticks, including harsh laws against the activities of a wide spectrum of potential mischief-makers such as perceived dissidents, terrorists, and protesters or rioters.[38]

At an operational level, Beijing must worry about maintaining the integrity of its strategic mobilization system, which is designed to ensure that the correct PLA units are deployed to the correct places at the correct times with the correct orders. This means the strategic mobilization system must enable real-time and uninhibited communication between units and their superiors. The mobilization decisionmaking system is considered the "nervous system" of the PLA and encompasses active-duty, reserve, militia, and even civilian enablers of military operations.[39]

Another operational influence point is the CCP's desire to preserve PLASSF capabilities, whether in peacetime, crisis, or wartime scenarios. As described above, the PLASSF is a relatively new service (as of 2016) responsible for collecting, analyzing, and disseminating information to the other four services. PLAASF focuses on cyberspace, space, and electromagnetic information. Without PLASSF, the rest of the PLA would be less informed prior to and during joint operations.

Chinese leaders also seek to maintain the capability to hold adversary national critical functions and infrastructure at risk. This is a broad operational objective that could include the use of offensive cyberspace capabilities, special forces (sabotage and other covert missions), or even conventional strike capabilities (namely, missiles). A 2022 think tank report outlines examples of potential Chinese counterforce attacks on commercial enablers of military operations, including core telecommunications systems, private logistics companies, off-base electric power sources, under-

---

[38] Willy Wo-Lap Lam, "'Stability Maintenance' Gets a Major Boost at the National People's Congress," *Jamestown Foundation China Brief*, Vol. 19, No. 6, March 22, 2019.

[39] Dean Cheng, "Converting the Potential to the Actual: Chinese Mobilization Policies and Planning," in eds. Andrew Scobell, Arthur S. Ding, Phillip C. Saunders, and Scott W. Harold, *The People's Liberation Army and Contingency Planning in China*, Washington, D.C.: National Defense University Press, 2015, pp. 120, 127–130.

sea cables, commercial satellites, and cloud services.[40] Of course, Chinese counterforce attacks could focus on strictly U.S. military targets as well, but Beijing might calculate that commercial targets are less likely to escalate the U.S. response. The think tank report also posits potential Chinese countervalue strikes, which might involve disruptions of the U.S. power grid, financial sector, health care systems, emergency services, telecommunications, or transportation networks.[41]

Beijing further seeks to defend theater command headquarters and command posts. Starting in 2016, Xi converted Beijing's original military region structure into five theater commands (TCs)—the Eastern, Western, Southern, Northern, and Central TCs. Their defining characteristic is that they are encouraged to operate jointly within their respective areas of responsibility (AORs) to carry out a wide range of defensive and offensive missions.[42] Under the new structure, TC commanders are probably empowered to do more with less oversight and need for directions from the CMC.

Related to conducting successful joint military operations is the need to preserve deployable major force groups to seize and maintain the military initiative. Beijing often speaks about the need to "seize the initiative" (i.e., act first in order to gain operational advantages, just prior to armed conflict).[43] Major force groups could include units with various roles and missions, but the common theme is that these units are available and prepared to successfully deny the U.S. military any early or easy victories that might demoralize or hobble future PLA joint operations.

Beijing will further seek to delay adversary contingency deployment to the region. In this area, the PLA probably takes it as a given that the U.S. mili-

---

40 "Preventing Chinese Sabotage in Crisis," in Jon Bateman, *U.S.-China Technological "Decoupling:" A Strategy and Policy Framework*, Washington, D.C.: Carnegie Endowment for International Peace, April 25, 2022, pp. 73–74.

41 Bateman, 2022, p. 74.

42 Edmund J. Burke and Arthur Chan, "Chapter 6: Coming to a (New) Theater Near You: Command, Control, and Forces," in Phillip C. Saunders, Arthur S. Ding, Andrew Scobell, Andrew N. D. Yang, and Joel Wuthnow, eds., *Chairman Xi Remakes the PLA: Assessing Chinese Military Reforms*, Washington, D.C.: National Defense University Press, 2019.

43 Daniel Shats, "Chinese Views of Effective Control: Theory and Action," China Aerospace Studies Institute, September 2022.

tary will intervene in a Taiwan or East China Sea scenario and perhaps in a South China Sea or India border scenario as well. Because of the emphasis Beijing places on "reunification" with Taiwan, this is where the PLA is furthest along in developing "active defense" concepts and operations. Active defense is forward oriented and uses positional and mobile warfare tactics in a layered defense network—essentially an anti-access/area denial (A2/AD) strategy—to prevent the enemy from attacking mainland China or interfering with PLA operations against an opponent.[44] Active defense requires the ability to conduct joint operations and power projection capabilities to attack an incoming enemy from afar. These are clearly the directions China is currently moving in. Within a naval and air warfighting context, active defense, as evidenced starting with the 2015 and later 2019 defense white papers, has expanded its range. No longer does active defense apply merely to "near seas," presumably the areas around Taiwan and perhaps the West Pacific and South China Sea, but also now "far seas" encompassing sea lines of communication as well as overseas economic interests. U.S. naval war planners should interpret active defense to be synonymous with Chinese counter-intervention capabilities, which at present are quite effective out to the First Island Chain; increasingly effective, though still in the early stages of development, out to the Second Island Chain; and aspirational beyond that range. Counter-intervention is primarily premised on launching ballistic and cruise missiles—for example, the DF-21D "carrier killer," the DF-26 "Guam killer," or the CJ-10 land-attack cruise missile—but also could include precision-guided munitions delivered by maritime strike and bomber sorties.

China's longer-ranging active defense network is designed to achieve another operational objective, which is to hold adversary forward-deployed forces, facilities, and materiel at risk. According to the latest DoD assessment, "the PLARF is developing and testing several new variants of theater-range missiles and developing capabilities and methods to counter adversary ballistic missile defense systems."[45] DoD goes on to state that "in 2020, a PRC-based military expert described the primary purpose of the DF-17 [medium-range ballistic missile] as striking foreign military bases

[44] M. Taylor Fravel, *Active Defense: China's Military Strategy Since 1949*, Princeton, N.J.: Princeton University Press, 2019.

[45] DoD, November 29, 2022, pp. 64–65.

and fleets in the Western Pacific." Because of the increasing threat of ballistic and cruise missile attacks, the U.S. military no longer has sanctuary in forward-deployed positions within the First and Second Island Chains, like Okinawa and Guam, respectively. Rather, the United States will have to harden existing facilities and disperse to new positions while maintaining its lethality to complicate Chinese targeting efforts.[46]

And finally, Beijing seeks to deny regional access to adversary expeditionary forces. This is another operational goal that nests within active defense because it involves leveraging the PLA's available forces to complicate the U.S. military's movement into and within the Indo-Pacific region.

Applying the Approach to Achieving Desired Effects Against an Influence Point: The Great Firewall as an Example

Information technology (IT) provides the primary means of information dissemination to and within the PRC. Therefore, IT also constitutes the primary infrastructure the CCP uses to prevent the free flow of information across its digital borders, censor information circulating within the GFW, and flood the PRC information environment with party-curated information. This information control system-of-systems is a double-edged sword. On the one hand, pervasive information controls provide the CCP with very powerful tools for communicating its edicts and manipulating the information environment. On the other hand, the party's dependence on this constant state of information control means that any potential outbreak of information and narratives incongruent with party intent presents a potential pitfall for CCP rule, and the CCP must scramble to quickly identify and quash such "rumors."

For assessing information-based influence points in the PRC, it is important to understand that the CCP's core strategies for information control predate the establishment of the PRC itself. Indeed, the first major intraparty struggle over rules of discourse and the boundaries of dissent

[46] David Ochmanek, *Restoring the Power Projection Capabilities of the U.S. Armed Forces*, RAND Corporation, CT-464, February 16, 2017.

occurred in 1942, during a seminal event known as the Yan'an Rectification. During this political inflection point, Mao and his close supporters faced a large contingent of CCP cadres and intellectual leaders who called for social, economic, and political reform. Rather than address the demands for reform, Mao and his supporters countered that the party was infallible;[47] that no type of knowledge besides "national struggle" could exist;[48] and that there was no abstract love, freedom, truth, or human nature. As a result, the only valid role for authors or artists within the CCP's area of control would be to reflect class difference and class struggle, in support of the CCP's ideological and political missions. The CCP then outmaneuvered its intellectual opponents, pressuring them one at a time to change their positions and eventually executing the last dissenters.[49]

The strategies the CCP used in 1942 were crude, and they focused on a relatively small population. During the Yan'an years, the CCP had lost control of China, and only a minority of the population was literate. However, the CCP's improvised survival strategies during the Yan'an Rectification became the foundation of its information control strategies after the CCP gained control over the remainder of mainland China. The same techniques served the CCP well when repressing intellectual dissent movements of 1957, 1976, and well into the information age. In fact, the CCP's current internet-based strategies for incentivizing PRC netizen self-censorship follow the same game theory patterns as the techniques first deployed at Yan'an in 1942.[50]

While the opening of China, initiated by then paramount leader Deng Xiaoping in the 1980s, brought dramatic economic growth to China, it also posed new and serious challenges to CCP legitimacy and PRC stability. The Tiananmen Square protests and subsequent killings of pro-democracy demonstrators in 1989 highlighted internal threats to the legitimacy of the CCP's continuing rule. The nascent IT revolution, a concurrent phenomenon that

---

[47] Anthony Saich, ed., *The Rise to Power of the Communist Party: Documents and Analysis*. Armonk, N.Y.: M. E. Sharpe, 1996. p. 1003.

[48] Saich, 1996, p. 1063.

[49] Anthony Saich and David E. Apter, *Revolutionary Discourse in Mao's Republic*, Cambridge, Mass.: Harvard University Press, 1994, pp. 66–67.

[50] Lev Navarre Chao, *Information Control Strategies of the Chinese Communist Party*, master's thesis, Harvard University Graduate School of Arts and Sciences, 2020.

grew exponentially in the 1990s and 2000s, presented further challenges to the CCP's ability to control information and achieve a stable environment for its continued rule.

Instead of preventing Chinese citizens from using the internet and social media, the CCP created what is referred to in the West as the "Great Firewall," to separate and isolate internet usage and information flow within China from the rest of the world. It also created alternative social media platforms, developed and owned by Chinese technology companies, such as WeChat and TikTok, while also developing a range of control mechanisms to ensure content created on those platforms deemed undesirable to CCP rule would be identified, minimized, or drowned out. With the creation and evolution of numerous and varied social media platforms, this ability to manage but not fully control information is no small task.

Unlike the development of the internet and social media systems in the West, which largely evolved in the creative commons and were driven by market forces in high-literacy societies, PRC systems of information-sharing and social media evolved under the guidance of the CCP and were pruned as the Chinese government saw fit to maintain control over citizens' access to information and citizens' capacity to share information with each other.

It is therefore important to note that the CCP's information control system-of-systems is not a bolt-on apparatus installed on top of an otherwise free information environment; it is baked in. For PRC citizens within the GFW, the very nature of information and truth was shaped and constrained by the CCP before the creation of modern China, before majority literacy and common spoken-language fluency among PRC citizens,[51] and before the existence of digital technology. There has never existed a time in modern history in which a majority of PRC citizens have both (1) been able to read and (2) had unfettered access to information. This has important

[51] China has historically been fractured by several hundred dialects, and Mandarin (the spoken-language lingua franca of China) did not pass the 50-percent prevalence mark among PRC citizens until the 20th century. See "More Than Half of Chinese Can Speak Mandarin," Xinhua News Agency, March 7, 2007. Fundamental literacy in the PRC (defined as recognition of minimum 1,500 written words for rural residents or 2,000 words for urban residents, insufficient to read a modern newspaper article) was only 57 percent as late as 1959. See Judy Heflin, "The Single Greatest Educational Effort in Human History," *Language Magazine*, undated.

implications for our understanding of influence points: there is no historical precedent for how PRC citizens might behave if they suddenly and en masse gained access to a free information environment.

The Great Firewall is really a misnomer, but it nevertheless is a useful term to refer to the numerous and varied organizations and functions that create and control information content and functions that design and regulate IT as well as provide defense, detection, and response to adverse information streams and actors.

Subsystems of the Great Firewall

What we call the Great Firewall is in fact officially referred to by the PRC government as the "Golden Shield Project" (金盾工程). From the outside of the "city"—that is, the PRC—we may only see a "wall," but from the inside of that city, the governing entity that is CCP views GFW as a shield, not a wall. In other words, we may see the GFW as a barrier to be breached—either by PRC citizens who wish to access the global internet or by the outside world wishing to share information with PRC citizens—but the CCP sees the GFW as a defensive tool for maintaining social stability among PRC citizens.

Furthermore, the GFW is only one tool in the CCP's toolkit, one system in the information control system-of-systems. For purposes of mapping PRC influence points and identifying potential targets and levers for escalation and de-escalation, it is better to understand the CCP's information control system holistically rather than simply focusing on the "wall" we observe from the outside.

There are several ministries and state instruments that hold joint and overlapping responsibilities for information control within the PRC.[52] The party's primary avenue for influencing information control is through the CCP Propaganda Department (中共中央宣传部). This is one of the oldest party organs; it was founded in 1924, suspended during the chaos of the Cultural Revolution, and reestablished in 1977 after Mao's death. It is responsible for oversight of both media censorship and control of state-sponsored information. At a more granular level, the State Council Information Office (国务院新闻办公室) and General Administration of Press and Publication

[52] Roberts, 2018.

(新闻出版总署) are responsible for published media, including the licensing of publishers and internet publishers, monitoring news and foreign journalists, and prescreening and banning books. Additionally, the arts and academia have dedicated information control systems. The Ministry of Education (教育部) is responsible for the regulation of information within the state-run system of elementary, middle and high school, and university education; there is very little access to private schools in the PRC except for noncitizens, and a 2021 regulation made private tutoring near-illegal.[53] The Ministry of Culture and Tourism (中华人民共和国文化和旅游部) has been responsible for regulation of the arts since it absorbed the Ministry of Culture (文化部) in 2018; the first minister of the newly created ministry was the former executive deputy director of the CCP Propaganda Department.[54]

The CCP has continued to evolve its approach to information controls by developing additional ministries as necessary to keep pace with the development of the internet and to guide that same development. The Ministry of Industry and Information Technology (工业和信息化部), established in 2008, is responsible for regulating the IT industry.

The State Council Information Office (SCIO, literally "State Council News Office") was originally created in 1991 as an external rebranding of the CCP Central Office of Foreign Propaganda (中央对外宣传小组) with the intention of mitigating damage to the CCP's reputation following the Tiananmen Square massacre.[55] The SCIO managed internet censorship in the PRC through its Internet Affairs Bureau until the 2011 creation of the State Internet Information Office (SIIO) (国家互联网信息办公室). SIIO remained subordinate to SCIO until it eventually grew large enough to warrant its own state nameplate.

---

[53] James Palmer, "Why China Is Cracking Down on Private Tutoring," *Foreign Policy*, July 28, 2021.

[54] 雒树刚被任命为首位文化和旅游部部长 [Luo Shugang was appointed as the first Minister of Culture and Tourism], *Economic Daily-China Economic Net*, March 19, 2018.

[55] Anne-Marie Brady, *Marketing Dictatorship: Propaganda and Thought Work in Contemporary China*, Lanham, Md.: Rowman & Littlefield, 2008, p. 23; and Anne-Marie Brady, "China's Foreign Propaganda Machine," Wilson Center, October 26, 2015.

In 2013, President Xi Jinping upgraded the SIIO to create a new and separate administration for regulating internet content and cyberspace, now called the Cyberspace Administration of China (CAC).[56] It was run by the Central Cybersecurity and Informatization Leading Small Group (中央网络安全和信息化领导小组) and personally chaired by Xi Jinping. In 2018, it was renamed the Central Cyberspace Affairs Commission (中央网络安全和信息化委员会) and is still chaired by Xi Jinping.

Not only can the government order traditional media to print particular articles and stories, but it also retains "flooding" power on the internet. The Chinese government allegedly hires thousands of online commentators to write pseudonymously at its direction.[57] CAC is also the majority owner of the China Internet Investment Fund, which has large stakes in social media companies such as ByteDance, which owns the social media platform TikTok.[58]

All these sprawling, often-overlapping, systems for controlling information flow across and within PRC together comprise parts of what we refer to as the GFW, but many of their functions exist within the PRC's information environment rather than at its perimeter, as the term "Great Firewall" would imply. These are not bolted-on systems repressing an otherwise free society; rather, they are integrated into the way the CCP has deliberately created an authoritarian state, and many PRC citizens may be unaware that any other global standard for interacting with information exists. Therefore, when considering these various segments of the CCP's information control system-of-systems for targeting purposes, we would be wise to frame any expectations for desired effects with the caveat that PRC citizens might not respond to GFW downtime the way we would expect citizens of liberal democratic countries to react. It is entirely possible that a PRC citizen who is unaware of the existence of Western media websites (e.g., Facebook, YouTube) might not attempt to take advantage of newfound internet

[56] Although the cyberspace administration is now rebranded in English as CAC, it retains the same Chinese office title of the SIIO (国家互联网信息办公室).

[57] These commentators are sometimes referred to as the 50 Cent Army or 50 Cent Party because they purportedly were paid half a renminbi (RMB¥0.50) for each post.

[58] Coco Feng, "Chinese Government Takes Minority Stake, Board Seat in TikTok Owner ByteDance's Main Domestic Subsidiary," *South China Morning Post*, August 17, 2021.

access to those sites during the downtime created by CYBERCOM missions to affect GFW influence points. To make any confident projections as to the actual felt effects on PRC citizens—and thereby project any secondary reactions by a CCP attempting to prevent, repair, or reverse damage to social stability—would require additional and in-depth research.

Potential Actions Against the Great Firewall Under Competition and Conflict

To continue this exploratory research and apply this report's framework, this section examines potential actions the United States could take against China's Great Firewall. Below we look at four ideal-type potential realities of competition and conflict, under low (minimalist) and severe (maximalist) conditions to further consider the possible cyberspace actions and intended effects as well as explore some considerations when carrying out such actions and possible ramifications.

Minimalist Competition

Under the conditions of minimalist competition, specifically a peacetime steady state that enjoys low levels of security competition between the United States and China, the United States would seek to conduct reconnaissance and surveillance of the Great Firewall.[59]

Reconnaissance would be necessary to identify and understand the unique topology of the organizations, entities, systems, networks, and functions that make up the Great Firewall. With this mapped out, the key nodes, their characteristics, and the connections between those nodes can then be used to begin to understand where potential vulnerabilities may lie.

Surveillance of day-to-day operations as well as crisis operations of the Great Firewall would also be necessary to see how the different aspects interact together, potentially exposing further seams and vulnerabilities and how the system responds to adverse information. Over even a nominal period, current events and the internal political dynamics of the PRC

[59] The degree to which these activities are conducted overtly will dictate whether they are considered minimalist actions or not. It is possible, however, that the CCP would see even minimal reconnaissance as aggressive or claim it is.

will provide plenty of case studies on how the Great Firewall operates and where inefficiencies and brittle points may lie. With this knowledge, concepts of operations and tools can be developed to exploit those specific vulnerabilities.

Maximalist Competition

Under the conditions of maximalist competition, specifically a peacetime steady state where severe security competition is occurring between the United States and China, the United States may seek to inject non-attributable sources of information into the Great Firewall or even degrade its operational effectiveness. During circumstances where competition may easily become conflict, these actions would seek to heighten the CCP's own concerns about its ability to maintain internal stability in order to reduce the likelihood of the PRC engaging in hostile actions against the United States or U.S. allies and partners that would likely invoke a U.S. response.

Of course, this action assumes that under such tensions, the PRC does not, contrary to the action's intent, lash out and initiate a conflict with the United States to mitigate its internal stability problems and deflect Chinese public attention by seeking to create a "rally around the flag" effect by stoking nationalist sentiment against the United States and its allies. Specifically, the action would attempt to inject information and narratives into the system damaging to CCP rule, amplifying adverse information that was internally produced by Chinese netizens while possibly degrading the Great Firewall's "antibodies" or effectiveness of internal reaction functions and their ability to respond to adverse information.

Minimalist Conflict

Under the conditions of minimalist conflict, the United States could seek to further degrade the functioning of the Great Firewall, increase access to outside content (e.g., *New York Times*, CNN, other websites), and reduce the Great Firewall's ability to track internal online behavior. This action is mostly an intensification of the maximalist competition, though in this case, U.S. cyber actions are less concerned about attribution or possibly even seeking to achieve attribution as kinetic conflict has already begun. As with maximalist competition, the intent and mechanism would be to demonstrate that the CCP's maintenance of internal stability is being jeopardized, and it should back off and terminate conflict or risk further internal instability.

Although certain action may meet the legal criteria of an armed attack and would create a de jure state of hostilities, it is possible to imagine scenarios of limited scope, duration, and intensity in which one or both sides in the conflict would wish to carefully calibrate armed responses in the expectation of containing the conflict. Such a hypothetical minimalist conflict could include instances in which the United States is supporting an ally or partner in responding to Chinese aggression that is not an existential challenge to the ally or partner. Second Thomas Shoal (the Philippines), Senkaku Islands (Japan), and the Sino-Indian border conflict (India) are potential flashpoints that fall into this category, assuming they do not escalate substantially. This category might also include certain instances where the United States and China are engaged in a hypothetical conflict that remains limited in scope and objectives, possibly over South China Sea right-of-passage issues, should the PRC seek to end U.S. freedom of navigation in this coastal sea. Minimalist conflict can escalate to maximalist conflict, and any cyber actions under this condition would need to be tempered so as not to cause such escalation.

Maximalist Conflict

Under the conditions of maximalist conflict, the United States could seek to fully disrupt the Great Firewall so that it ceases to effectively function in filtering and limiting access to content both internally and externally. The objective would be to allow Chinese netizens unmonitored and uncensored communication as well as unfiltered access to the internet. The intended outcome would be to pose severe challenges to the CCP's continued ability to govern the PRC. During maximalist conflict, this action may even entail kinetically targeting specific nodes within the Great Firewall to physically damage its ability to function. Furthermore, attribution would not be a concern. Actions toward this end, whether kinetic or nonkinetic, would need to be carried out in full accordance with U.S. strategic objectives because the escalation potential is severe, and forgoing such actions may provide a needed future "carrot" to de-escalate.

By its nature, maximalist conflict suggests that the strategic leadership in the United States and China would view themselves to be locked into an existential struggle, and the highest rung of the escalation ladder will have been reached. Yet there still may be some trade space within the sphere of hypothetical scenarios of existential conflict. From least to the most severe, existential conflict alternately could be based on both belligerents believing the outcome of the conflict will determine who controls the Asia-Pacific region, whether the U.S. government or CCP will remain in effective control of their populace, and whether American and Chinese society continues to function based on mutually assured nuclear annihilation through nuclear exchange. As a result, the United States may seek to forgo certain actions against the Great Firewall as a "carrot" or an "off ramp" in order de-escalate the existential crisis that maximalist conflict creates. In other words, the carrot of allowing the CCP to continue to remain in power is likely to be perceived in almost every case as a less bad outcome than a hypothetical nuclear exchange.

Complexity in Leveraging Chinese Influence Points, Especially the Great Firewall

The logic underlying the identification of leverage points is the matching of elements and objectives of the PRC system that are (or should be) important to the CCP with opportunities to affect them in ways that reduce PRC strength, posture, position, or the ability to realize objectives. Because the combined systems of Chinese governance and society are complex (and exist within the complexities of the global political, economic, and social context), complexity theory applies.

Complexity theory is premised on the idea of complex holism and is in part a backlash against reductionism in social science, the drive to understand an entity through reduction to its smallest parts.[60] Complexity theory involves a rejection of the idea that all unknowns are known and instead

[60] John R. Turner and Rose M. Baker, "Complexity Theory: An Overview with Potential Applications for the Social Sciences," *Systems*, Vol. 1, No. 1, 2019.

embraces dynamic adaptation of both the systems under consideration and what we know about them.[61]

Complexity theory includes several concepts with relevance to seeking and using selective overmatch to plan actions across the continuum of competition and conflict against Chinese leverage points. These include path dependency, sensitivity to initial conditions, nonlinearity, power law phenomena, holism and irreducibility, emergence, feedback and interdependency, contingency, and adaptiveness.[62]

What does this mean for the current analysis? Collectively, the various concepts of complexity theory add up to *increased uncertainty* about the effects of various actions against various leverage points. This could manifest as uncertainty about the amount of pressure required to cause some sort of change or effect—we may assume that a leverage point is vulnerable or brittle, but there may be unseen complex reinforcement mechanisms or institutional inertia built into various systems that create resilience. Or some structures may be more brittle than they appear, and light pressure may lead to cascading failures. This is an example of uncertainty about the scope and direction of consequences of actions—that is, there is a significant chance of unanticipated consequences, either because of poorly understood (or superficially incomprehensible) interrelationships between different systems or because of emergent or novel phenomena. There is also uncertainty about how the CCP and PLA will respond to various stimulations of leverage points; beyond the danger of mirror imaging (where U.S. planners fail to account for the fact that Chinese planners and strategists may have different

[61] Mario Couture, "Complexity and Chaos—State-of-the-Art: Overview of Theoretical Concepts," technical memorandum, Defence R&D Canada, August 2007.

[62] For more extensive discussion on these topics, see Turner and Baker, 2019; Couture, August 2007; Michael R. Weeks, "Chaos, Complexity and Conflict," *Air and Space Power Chronicles*, Vol. 6, 2001; Mary Lee Rhodes, and Elizabeth Eppel, "Public Administration and Complexity—Or How to Teach Things We Can't Predict?" *Complexity, Governance & Networks*, Vol. 4, No. 1, January 2018; Jun Yan, Lianyong Feng, Artem Denisov, Alina Steblyanskaya, and Jan-Pieter Oosterom, "Complexity Theory for the Modern Chinese Economy from an Information Entropy Perspective: Modeling of Economic Efficiency and Growth Potential," *PLoS ONE*, Vol. 15, No. 1, 2020; Andrew Bennett and Colin Elman, "Complex Causal Relations and Case Study Methods: The Example of Path Dependence," *Political Analysis*, Vol. 14, No. 3, Summer 2006.

preferences and understandings than what the U.S. would), complexity can filter perceptions based on information flows within a system or unpredictable outcomes in some systems. Bottom line: some actions against leverage point could be perceived as much more threatening than anticipated, so it is possible that actions intended as competition actions could be unexpectedly provocative or escalatory.

How should we think about this? We suggest two perspectives: First, if probability and outcomes are likely to fall within a confidence interval (where narrower intervals suggest greater precision in prediction, and broader intervals suggest less certainty): broaden the intervals. Our confidence about exactly what will happen should be admittedly lower than we would like and that traditional projections would suggest. Second, and related, the range of possible outcomes—perhaps in terms of a competitor and adversary course of action (COA), traditionally including the most dangerous COA and most likely COA—needs to be broader and should explicitly include an *unanticipated* COA or outcome. Certainly, planners should seek to imagine all conceivable possible outcomes, but where complexity obtains, sometimes the inconceivable occurs.

As noted, the GFW is not a bolt-on apparatus installed on top of an otherwise free information environment; it is baked in. And it has been baked in long enough that most citizens might be unaware of any other global standard for interacting with information. So, one outcome from short duration breaches in or weakening of the GFW could be that most Chinese citizens do not notice or do not know how to take advantage of the opportunity if they do notice. It is more likely that some members of the public will notice and take advantage of freer access to information, but the Ministry of Public Security is also likely to detect and respond to these breaches quickly. If there is a sustained outage coupled with an earnest effort (organic or external) to promote different modes of engagement with information, it is likely that things begin to change, but possible outcomes are subject to chaos and complexity. It is possible to partially enumerate these changes or outcomes, but such complex systems as a whole society with GFW as part of its social fabric could lead to emergent and totally unanticipated outcomes.

Extreme outcomes are possible. And because extreme outcomes are possible, the level of threat or aggression perceived by the CCP could be disproportionate to the intended level of aggression by the United States, *even if no significant changes result.* Thus, uncertainty about effects increases escalation risk. Obviously, if effects on the GFW resulted in significant social upheaval, it would be viewed as escalatory, but that might also be the case even if (unpredictable) consequences are minimal.

Conclusion

This chapter identifies key influence points within the PRC toward which U.S. cyber capabilities could be applied; and examines one of those influence points in greater detail, highlighting the Great Firewall's organization, structure, and functions while discussing general targeting approaches and actions to take during competition and conflict. The United States still needs to understand more about the Great Firewall to understand which nodes are most important and what will be effective. For example, the firewall might be down, but if PRC netizens are very nationalist at that period, it may not matter. As a result, the above analysis is exploratory rather than prescriptive. Significant additional research and analysis needs to be done to understand even basic causality within the Great Firewall and to what extent its key nodes are even targetable. At this early point in such research, it is difficult to state with any certainty what the effects would be, if any, of targeting any of the above-mentioned nodes or entities. We must also be cognizant that the very aspects that make the Great Firewall a potentially attractive target for U.S. CYBERCOM actions can also lead to unintended and even severe escalation, especially if the CCP sees itself, because of or in conjunction with these actions, losing its grip on control of the PRC.

It is therefore necessary for us to understand the limitations of targeting the GFW (and affiliated systems of information control) in terms of predicting the second- and third-order effects of successful missions to degrade or destroy these censorship apparatuses. For example, it is entirely possible that degrading the GFW during a time of organic PRC citizen mass dissent against the CCP would have a tremendous impact on social stability within the PRC, forcing the CCP to redirect resources inward and quell riots and citizen resistance. On the other hand, a comparable mission carried out during a time of elevated PRC citizen nationalism might have a negligible effect because citizens largely support the CCP and are not especially interested in coordinating with each other to riot or inclined to seek information from the outside world—even if they were aware of which URLs to enter into their browsers. As a result, much further analysis needs to be carried out on all the potential influence points, as well as numerous others not identified within this exploratory chapter.

Furthermore, even if the Great Firewall's information control functions are substantially disabled or destroyed, it is unclear that the opportunity could be successfully exploited. Even with China's information defenses down, to capitalize on an opportunity, the United States would need to insert alternative narratives that achieve some modicum of groundswell support in the popular imagination or adeptly identify and amplify otherwise quashed narratives originating from Chinese citizens. Because of the CCP's relative stranglehold on information-sharing and its multiple generations of experience successfully repressing dissent and degrading citizens' capabilities to coordinate with each other, many PRC citizens not only lack awareness of global standards for information availability and sharing, but they may in fact lack the wherewithal to take advantage of free access to the global internet even if granted access to it.

CHAPTER SIX

Defending U.S. Influence Points

In this chapter, we examine the other side of the coin: the potential influence points the PRC would seek to affect and how the United States, employing the concept of selective overmatch and the I2E2 capabilities and approaches, can deter threats to and defend those influence points.

Just as U.S. strategic competitors have influence points that the United States will seek to affect, so will our adversaries do the same. The United States' adversaries, particularly those dissatisfied with the current international order, will try to undermine, constrain, and influence U.S. decision-making. It is important, therefore, for the United States to understand what its influence points are and how to defend them.

International relations theorists generally recognize a few core national interests in any country: security, both from external attack and internal unrest; prosperity, with a sufficient level of present-day wealth and future economic growth to ensure the well-being of its citizens; and liberty or political stability, in which citizens of rule-of-law countries maintain core individual rights and their governments retain an appropriate separation of powers. In countries that have authoritarian or undemocratic rule, this interest takes the form of political stability under a ruling regime, and stability is measured in terms of its retention of power.

Some theorists argue there is a fourth national interest, usually referred to as value extension. This captures the overlap of one nation's interests with another's and the degree to which a country will commit wealth and accept risk in order to defend its core interests overseas. Value extension can take two forms: *isothymic*, which involves protecting the territorial integrity and political independence of other states as well as defending the fundamental human rights of the citizens of those states; and *megalothymic*, which is the imposition of a powerful country's will on a weaker neighbor without the consent of its government or its citizens, with an eye toward indefinite occupation and permanent territorial annexation.

There is a robust history of debate about what constitutes U.S. national interests and even whether the term itself is useful.[1] Scholars have argued that at the inception of the republic, the United States had two national interests: (1) a desire for a balance of power in Europe to ensure that no one power could threaten the survival of the young nation, and (2) freedom for trade.[2]

The extent to which morality (or value extension, as noted above) should and does play a role in defining national interests is also debated, with some arguing that realists (in the mold of Hans Morgenthau) are inherently amoral, while Morgenthau and Samuel Huntington argue that realism is fundamentally a moral position because it is focused on the survival of the state. Other theorists have articulated morality in U.S. national interests to mean the pursuit of ends such as human rights or the spread of democracy.

In 1996, the Commission on America's National Interests opined that the collapse of the Soviet Union and the lack of a unifying opponent had led to a rudderless foreign policy. The commission went on to define five "vital national interests":

- Prevent WMD attacks on the United States.
- Prevent the emergence of hostile powers in Europe, Asia, or on the United States' borders.

---

[1] There is considerable literature on the topic of the national interest, which we only touch on briefly here to establish the foundation for our subsequent development of influence points. For more on the debate, see Hans J. Morgenthau, "Another 'Great Debate': The National Interest of the United States," *American Political Science Review*, Vol. 46, No. 4, December 1952; Hans J. Morgenthau, "What Is the National Interest of the United States?" *Annals of the American Academy of Political and Social Science*, Vol. 282, July 1952; Anthony Lake, "Defining the National Interest," *Proceedings of the Academy of Political Science*, Vol. 34, No. 2, 1981; Michael G. Roskin, *National Interest: From Abstraction to Strategy*, Carlisle, Penn.: Strategic Studies Institute, U.S. Army War College, 1994; Samuel P. Huntington, "The Erosion of American National Interests," *Foreign Affairs*, Vol. 76, No. 5, September–October 1997; and Joseph S. Nye, Jr., "Redefining the National Interest," *Foreign Affairs*, Vol. 78, No. 4, July–August 1999.

[2] Hans J. Morgenthau, "The Policy of the USA," *Political Quarterly*, Vol. 22, No.1, 1951; and Adam Lowther and Casey Lucious, "A Call for Action: Defining the U.S. National Interest," *Atlantisch Perspectief*, Vol. 7, No. 37, 2013, p. 5.

- Prevent the emergence of a hostile major power on U.S. borders or in control of the seas.
- Prevent the collapse of global systems for trade, energy, financial systems, and the environment.
- Ensure the survival of U.S. allies.[3]

These are interestingly all essentially status quo and "negative" interests in the sense that they seek to prevent adverse action from undermining an international system that is conducive to continued United States' integrity, survival, and prosperity. In 2000, the commission modified these national interests, placing the survival of U.S. allies second in its list and adding the need "to establish productive relations, consistent with American national interests, with nations that could become strategic adversaries, China and Russia."[4] Joseph Nye noted that "as a wealthy status quo power, the United States has an interest in maintaining international order."[5]

Based on a review of academic and public policy literature, the meaning of "the national interest" and what that encompasses remains deeply contested.[6] Writing in the 1970s, Donald Nuechterlein postulated three primary national interests:

- protection of the people, territory, and institutions of the United States against potential foreign dangers
- promotion of U.S. international trade and investment, including protection of U.S. private interests in foreign countries

[3] Quoted in Huntington, September–October 1997, p. 36; and Robert Ellsworth, Andrew Goodpaster, and Rita Hauser, *America's National Interests: A Report from the Commission on America's National Interests*, Cambridge, Mass.: Belfer Center for Science and International Affairs, Harvard Kennedy School, July 1996, pp. 2–3. This 1996 report identified the need to manage the rise of China and try to prevent the resurgence of authoritarianism in Russia or its collapse.

[4] Robert Ellsworth, Andrew Goodpaster, and Rita Hauser, *America's National Interests: A Report from the Commission on America's National Interests*, Cambridge, Mass.: Belfer Center for Science and International Affairs, Harvard Kennedy School, July 2000, p. 3.

[5] Ney, July–August 1999, p. 27.

[6] See Donald E. Nuechterlein, "The Concept of National Interest," in *United States National Interests in a Changing World*, Louisville: University Press of Kentucky, 1973. This work has a useful overview of the major writers on the topic.

- establishment of a peaceful international environment in which disputes between nations can be resolved without resorting to war, and in which collective security rather than unilateral action is employed to deter or cope with aggression.[7]

Nuechterlein noted that the first two interests have endured since the start of the republic, while the last was of more recent advent, particularly at the end of World War II. He includes promotion of democracy and free enterprise abroad as a part of the third interest, though others have argued for this as a separate national interest.

Our purpose in reviewing the literature on national interests is to demonstrate that while not entirely settled, the concept of territorial integrity, economic prosperity, and some form of international order that brings stability and predictability are generally (though not universally) seen as core interests.[8] The relative balance across these interests may shift from administration to administration and based on perceptions of threats and opportunities in the international environment. To guide U.S. actions and planning, however, it is necessary to develop the influence points that underpin these national interests. We use these national interests to now disaggregate the national interests into policy, strategic, and operational influence points that the United States should seek to protect and that we would expect adversaries to try to undermine or affect to achieve their goals. We accomplished this through a series of internal RAND discussions with subject-matter experts and two rounds of refinement to the list developed in those discussions.

Policy-Level Influence Points

U.S. national interests, as noted above, are contested in how broadly they should be defined. The Commission on America's National Interests argued that the vital national interests should adhere to the common dictionary

7 Nuectherlein, 1973, p. 8.

8 Many writers will make distinctions between "vital," "essential," "important," and "secondary" national interests. For our purposes, we do not address these distinctions.

definition of "vital": "conditions that are strictly necessary to safeguard the well-being of Americans in a free and secure nation."[9] We identified four influence points that, if affected, could undermine U.S. national interests. We call these policy-level influence points.

Security of the United States. To secure the United States' territory, protect against external threats, and sustain the United States as a state requires providing for the common defense against external threats and ensuring domestic tranquility by addressing internal threats. An adversary could seek to affect the United States by posing a threat from abroad to U.S. territory through physical attack or by undermining and exploiting fissures in domestic stability.

Prosperity. The United States derives its power through its economic might, which depends on robust economic activity domestically and access to world markets. An adversary could seek to undermine the United States economically.

Liberty. Individual liberty to pursue social and economic well-being is supported by a constitutional government that protects minority rights and mediates disputes through judicial and political processes rather than violence; that is, it requires stable governance and continued agreement from the governed to abide by decisions made through political processes. An adversary could seek to undermine this social compact to sow further division in the United States and convince significant numbers of Americans that violence or other non-democratic means are legitimate ways to act.

Extended values. The United States sees its values of democratic governance and individual liberty as universal values that support security and prosperity domestically and abroad. An adversary could seek to undermine the acceptance of these values internationally or present an alternative set of values, similar to what the Soviet Union did during the Cold War.

Strategic-Level Influence Points

Strategic-level influence points are more tangible national-level capabilities and goals that underpin the preservation of the policy-level influence

[9] Ellsworth, Goodpaster, and Hauser, 1996, p. 4.

points. For each influence point, we provide a brief description and explain why an adversary might choose to try to affect it.

Ensure security, reliability, and lethality of the strategic triad. The strategic triad—the combination of submarine-based nuclear missiles, land-based nuclear missiles, and manned nuclear bombers—is the ultimate guarantor of U.S. security providing a deterrent to existential threats to the United States from powers like Russia. It also serves a role in preserving the territorial integrity and security of our closest allies. An adversary could seek to undermine U.S. confidence in the effectiveness of the triad.

Sustain primacy of U.S. and allied military power. Military power deters conventional threats to territorial integrity and security. Military power also contributes to securing the global commons, such as the high seas to support the free movement of people and goods that contribute to prosperity. An adversary could seek to degrade military power directly through conflict or indirectly by undermining confidence in the effectiveness of military power.

Deter adversaries from aggression against U.S. and allies. Aggression against the United States or its allies threatens territorial integrity and security. By deterring aggressive action, the United States continues to secure itself and its allies and signal that they will not tolerate destabilizing military and other coercive action. An adversary could carry out military or other coercive actions to undermine faith in U.S. support to allies' territorial integrity and sovereignty.

Ensure common domains (sea, air, space, cyberspace) remain open and free. The common domains support economic activity and the exchange of people, goods, and ideas that contribute to U.S. security and extend values that are seen as conducive to sustaining that security. Constraining access to those domains can undermine U.S. economic activity and prevent the free movement of people, goods, and ideas.

Serve as most capable and reliable security partner of choice. Sustaining its preeminent security and military capabilities demonstrates the United States' commitment to the international rules-based order and its allies' territorial integrity and security. Similar to deterring aggression against the U.S. and its allies, an adversary could carry out actions to undermine allies' faith in this commitment.

Preserve current rules-based order domestically and overseas. The current international rules-based order serves U.S. national interests well. It sustains U.S. economic prosperity and leadership on the international stage while establishing norms of acceptable international behavior. An adversary could try to undermine this order because it sees the current system as biased toward the status quo powers and not according respect and opportunity for growth to the adversary.

Maintain faith in legitimacy and effectiveness of governmental institutions. The United States and other democracies are based on the consent of the governed. Despite declining trust in public institutions and the growth in anti-government groups, most Americans still believe in democratic institutions.[10] Extremism has grown faster in online forums compared with instances of violence in public spaces.[11] This faith in the legitimacy of governmental institutions facilitates political processes to determine how to run the country and make decisions about U.S. action on the international stage. An adversary could try to undermine this faith to delegitimize government's decisions and policies in the eyes of the populace and, in extreme circumstances, produce violent insurrection.

Protect integrity and reliability of national critical functions and infrastructures. National critical functions are "the functions of government and the private sector so vital to the United States that their disruption, corruption, or dysfunction would have a debilitating effect on security, national economic security, national public health or safety, or any combination thereof."[12]

Sustain and advance U.S. competitive edge in education, research, and development. The United States is recognized globally for the quality of its colleges and universities, both in terms of education and research. Additionally, the United States continues to improve its productivity and thereby

---

[10] For the decline in public trust in government institutions, see Pew Research Center, "Public Trust in Government: 1958–2022," June 6, 2022.

[11] For the growth in anti-government groups, see Heather J. Williams, Luke J. Matthews, Pauline Moore, Matthew A. DeNardo, James V. Marrone, Brian A. Jackson, William Marcellino, and Todd C. Helmus, *A Dangerous Web: Mapping Racially and Ethnically Motivated Violent Extremism*, Santa Monica, Calif.: RAND Corporation, RB-A1841-1, 2022.

[12] Cybersecurity and Infrastructure Security Agency, "National Critical Functions Set," website, undated.

grow its economy through innovation and the development of new technologies. Undermining this influence point could reduce the quality and quantity of highly skilled and educated workers and lead to others surpassing the United States in developing new technologies.

Sustain and advance U.S. commercial advantage in the development, production, and distribution of goods and services. The United States is predominantly a service economy, but it still has significant production capacity.[13] An adversary seeking to weaken the U.S. economy could engage in industrial espionage to steal trade secrets and copy technologies.

There are many ways an adversary could try to undermine or weaken the United States by affecting one or more of the strategic influence points. Through these actions, adversaries would attempt to attack U.S. national interests and the policy-level influence points. While the completeness of this list is subject to debate, it represents a cross section of areas that we assess an adversary would examine to determine how to challenge the United States' primacy.

Operational-Level Influence Points

The strategic-level influence points we developed in the previous section directly underpin the four policy-level influence points. Strategic-level influence points are supported in turn by influence points at the operational level. These operational influence points were similarly developed and refined through internal discussions with RAND subject-matter experts.

Defend the homeland from strategic or conventional attack. The DoD maintains numerous capabilities to defend the homeland from strategic or conventional attack, including missile defense systems and the CMF's NMTs. An adversary seeking to overcome these defenses would target them directly or indirectly.

Minimize the proliferation of WMD and related delivery technologies. Global security and prosperity would be severely disrupted by the prolifera-

[13] Bureau of Economic Activity, "Gross Domestic Product," website. The bureau releases monthly estimates of gross domestic product (GDP), corporate profits, and GDP by industry.

tion of weapons of mass destruction (WMD) and related delivery technologies. This proliferation would allow disruptive regimes and nonstate actors to threaten their neighbors, destabilize critical regions, and increase fear of attack that could inhibit the free flow of goods, services, and people. An adversary could deliberately support proliferation to cause these adverse effects.

Preserve the ability to securely project decisive military power. The U.S. military operates globally and relies on the ability to project decisive military power from the homeland and bases in regions around the world. This ability is critical to supporting allies and partners, defending U.S. interests, and deterring aggression. An adversary could seek to prevent the United States from engaging in a conflict or crisis with its military capabilities by denying the ability to project that power.

Preserve the ability to sustain forward-deployed and expeditionary forces. Following on the ability to project power, the United States has to sustain the forces it has in regions like the East Asia and Southwest Asia. The United States also responds to emerging crises and conflict through the use of expeditionary forces that can establish and sustain a presence and conduct operations in austere environments. Without these capabilities, the United States would not be able to respond quickly to address aggression or deter an adversary; instead, it would be faced, in some cases, with a fait accompli by the time it could bring military power to bear.

Maintain the ability to supply allied and partner nations under attack. As the United States and its allies have demonstrated in the Ukraine conflict, supplying partners with military and other supplies during conflict is an important component of addressing the actions of revisionist powers.[14] Without this ability, allies and partners would succumb to aggression that would jeopardize their territorial integrity and sovereignty.

Maintain a strategic global sensor and communications grid. The United States can only defend against threats such as a strategic attack and project power through its vast capabilities to monitor and communicate globally.

---

[14] John Masters and Will Merrow, "How Much Aid Has the U.S. Sent Ukraine? Here Are Six Charts," Council on Foreign Relations, December 16, 2022.

An adversary could take actions to blind or deceive the United States at critical moments in a crisis.

Preserve the ability to establish and maintain operational communication and sensor grids when and where needed. In addition to maintaining the strategic sensor and communications grid, the United States needs the ability to establish and maintain operational-level capabilities to maneuver forces and monitor adversary actions across domains. An adversary can seek to deny the United States the ability to command and control its forces in a theater of operations.

Protect supply chains and access to markets for critical goods and services. U.S. economic prosperity relies on access to critical inputs through global supply chains and access to markets to sell its goods and services. As we saw during the coronavirus disease 2019 (COVID-19) pandemic, disruptions to global supply chains can persist for years. The closing of markets would also negatively affect the United States.

Ensure the security and continuity of the global energy distribution system. Economic activity and military power both rely on access to energy in various forms, from liquid natural gas to nuclear power and renewable energy sources. Experiences like the oil embargo in the 1970s and the ransomware attack on the Colonial Pipeline in 2021 demonstrate how disruptive adverse actions can be to the distribution of critical energy.

Protect and advance the legitimacy of international institutions. International institutions like the World Bank and NATO have shaped the post–World War II order and sustained U.S. primacy for more than 75 years. These international institutions contribute to regional and international stability and support security and economic prosperity. Undermining international institutions could undermine the functioning of the international rules-based order.

Preserve the integrity of the global financial system, access to credit, and monetary stability. The global financial system sustains economic activity and U.S. prosperity. The United States also enjoys a privileged position and power because of the central role its currency plays. Economic growth is fueled by investment stemming from access to credit. Disruptions to these systems could wreak havoc on the international system.

As with the strategic-level influence points, an adversary could try to undermine U.S. national interests through numerous effects on one or more of the operational-level influence points, whether to prevent the United States from intervening in an emerging crisis or conflict or undermining U.S. prosperity.

Applying the Approach to Analysis of an Influence Point: Holding Adversary Forward-Deployed Forces, Facilities, and Materiel at Risk

We now turn to the application of strategic overmatch and the I2E2 construct to a Blue-focused effort, defending against an adversary's attempts to deny the United States an operational advantage through the use and sustainment of forward-deployed forces in crisis and conflict. Following the format established in Chapter 5, we start by looking at potential actions the United States would employ across the four quadrants of competition and conflict. This influence point is written from the adversary's perspective—that is, the adversary would seek to hold the United States' forward-deployed forces, facilities, and materiel at risk. The United States would try to defend and thwart the adversary.

Minimalist Competition

Under conditions of minimalist competition, the United States objectives would be to prevent the PRC from developing the posture and capabilities to hold our forward-deployed forces, facilities, and materiel at risk. The United States would not expect overt hostile action from the PRC against U.S. military forces, although it has engaged in harassing actions such as a 2009 incident involving a U.S. surveillance vessel navigating in international waters.[15] PRC cyberspace actors could engage in propaganda efforts against the United States and may seek to probe systems and networks for intelligence purposes or for operational preparation of the environment. To counter this, CYBERCOM could conduct hunt forward missions focusing

[15] David Morgan, "U.S. Says Chinese Vessels Harassed Navy Ship," *Reuters*, March 9, 2009.

on systems that support U.S. power projection into and intra theater; conduct cyber risk assessments of major systems and networks; and conduct cyberspace intelligence, surveillance, and reconnaissance (ISR) to identify potential vulnerabilities in adversary (red) systems that threaten U.S. power projection.

With the development of I2E2 capabilities and approaches, CYBERCOM can contribute to counterintelligence and hunt operations to dislodge adversary cyber actors from key systems and networks that support United States Indo-Pacific Command (INDOPACOM) contingency plans and United States Transportation Command's ability to deploy in response to contingencies. The United States would also conduct collaborative cyber risk assessments with partners and allies in the region to extend these efforts to the critical systems they operate. CYBERCOM can also develop integrated "flexible deterrent options" that employ capabilities across domains to respond to rising tensions or early indications of potential future Chinese aggression.[16] Cyber resiliency for forward-deployed forces, facilities, and materiel is also a critical component of deterring aggression and ensuring U.S. freedom of action. Finally, CYBERCOM can work with partners and allies to conduct collaborative hunt forward missions to identify potential infrastructure to monitor for indications of future adversary cyberspace operations, and it can work with cyberspace ISR to develop greater understanding of Chinese A2/AD capabilities and systems, which can identify potential vulnerabilities and support target planning.

Maximalist Competition

Maximalist competition could be characterized by more aggressive Chinese provocative action or by U.S. action to push back more assertively against Chinese actions. U.S. cyberspace activities could encompass more aggressive hunt forward missions, signaling actions through cyberspace and other means to notify the Chinese that the United States is aware of

[16] Flexible deterrent options are intended to provide the ability to employ military forces in a variety of ways to control escalation in a conflict. They were first developed as flexible response options in nuclear and deterrence doctrine.

and can counter Chinese cyberspace activity that threatens U.S. freedom of action. This signaling could include targeted actions to identify and remove adversary cyberspace actors and actions to indicate the ability to hold Chinese forces at risk. These actions are all within CYBERCOM's current capabilities.

Employing additional capabilities and approaches as laid out in our I2E2 construct could involve conducting an influence campaign integrating cyberspace, information, deception, and intelligence to undermine PRC confidence in its power project capabilities. By coordinating the campaign across multiple U.S. agencies and select partners and allies to integrate effects across the physical and cyber domains, the United States would not only demonstrate technical capability to counter Chinese actions, but it would also sow doubt in key decisionmakers' minds that they can prevent the United States from intervening in an emerging crisis. An important component of this approach (stemming from the "integral cyber" capability area) would be also integrating cyberspace "signals management" in concert with deception to reduce the adversary's capability to accurately track U.S. forces in real time. Combined and joint exercises with regional partners can also be used to demonstrate hardening and resiliency of our forward-deployed forces and installations. At the same time, cyberspace ISR, whether used unilaterally or combined with partners and allies, can identify vulnerabilities in China's A2/AD capabilities and support development of plans and capabilities to counter them. Currently, CYBERCOM's capabilities are limited to carry out these types of activities.

Minimalist Conflict

In a conflict with China, the President and the Secretary of Defense will want to have options available to them across domains to counter any Chinese aggression while controlling escalation. Targeting certain systems and nodes, while operationally desirable, could have strategic disadvantages, particularly if it leads to escalation on China's part or their perception that U.S. objectives are aimed at undermining regime stability. To sustain the ability to project power and support forward-deployed U.S. forces, the United States can undertake a number of actions in cyberspace. Given the current construct and posture of the CMF, a focus for CYBERCOM could be on conducting cyberspace operations to disrupt or degrade China's mobilization of forces and command and control of those forces.[17]

With the addition of I2E2 capabilities and approaches, additional actions in minimalist competition could encompass integrated operations, including cyberspace, EW, and kinetic actions to blunt China's ability to threaten U.S. power projection and forward-deployed forces, in particular the key systems identified in Chapter 5, such as the DF-21 and CJ-10. Additionally, cyberspace ISR can play an important role in identifying threats and providing tactical warning to forward-deployed forces of those threats and defensive cyberspace operations (DCOs) to protect critical systems at forward-deployed stations. CYBERCOM currently has nascent access to additional I2E2 capabilities, but not to the extent that would be required to implement this new approach successfully at the scale required.

Maximalist Conflict

In maximalist conflict, the United States is still seeking to control escalation where possible, but it also recognizes that a broader array of targets and actions may be required to respond to Chinese aggression. At a minimum, CYBERCOM could conduct cyberspace operations targeting power projection forces and supporting infrastructure, including electric power generation and distribution, to degrade China's ability to threaten forward-deployed forces. These operations could also extend to undermining the PLA's confidence in key systems such as the DF-21. CYBERCOM currently has many of these capabilities.

When employing additional I2E2 capabilities and approaches, integrated operations would extend to attacking a broader array of targets, including supporting infrastructure and systems (e.g., logistics, maintenance systems). It would include an extended information campaign to underscore PRC aggression and provide support to opposition groups and individuals within China. DCOs would be integrated with operational planning to

[17] It is important to note that this statement is not addressing any current CYBERCOM plans or extant capabilities. Rather, it is a statement about the general focus and posture of cyberspace forces and the types of targets and actions they could be employed against.

provide enhanced protection for platforms and systems critical to military response and with cyberspace ISR to provide tactical warning of threats to U.S. forward-deployed forces and systems. Cyberspace operations as part of a broader deception campaign can also serve to improve operational security for U.S. forces and mislead Chinese sensors as to U.S. operating locations and intentions. CYBERCOM's integration with other information-related capabilities and deception is limited currently.

Complexity and Holding Adversary Forward-Deployed Forces, Facilities, and Materiel at Risk

There are multiple levels of complexity in the cat-and-mouse game where the PLA seeks to hold U.S. forward-deployed capabilities at risk and the U.S. seeks to counter those efforts. The first level involves the underlying and interconnected systems of international relations and international economics for the two antagonists and relevant regional neighbors, and the second level involves the competing systems of military capabilities (i.e., the forward-deployed capabilities; the capabilities that might threaten those capabilities; and the capabilities that might counter, mitigate, or reduce that threat).

The higher level of complexity (international relations and interconnections) holds the greatest potential unpredictability in this case: how potential partners (and their citizen constituencies) will react when their sovereign territory becomes the operational environment for vigorous competition or conflict; economic knock-on effects of investment by China, the United States, or both; unpredictable reputational effects from either principal's posturing or actions; and so forth. That said, though governed by complexity and including many unpredictable effects, outcomes at a still higher level are fairly constrained. That is, affected economies and firms will either benefit or suffer; and potential partners will either align with the United States to some extent, align with the PRC to some extent, balance carefully between the two, or take another of the (known and constrained) options available to a country in its international affairs.

Similarly, at the lower level of complexity, the complex interplay between the military systems interacting in the cycle of adaptation and counteradaptation to deny or preserve U.S. forward presence governs the outcomes of those interactions, but the range of possible outcomes is reasonably constrained. In any given cycle and for any given capability, either the capability is held at risk to some significant extent, or it is predominantly safe. The outcome is causally complex, but there are basically only two outcomes (or partial versions of those outcomes).[18]

Cumulative complexity across both levels holds considerable uncertainty but culminates in narrower strategic outcomes: to some extent, either U.S. forward-deployed capabilities continue to serve their deterrent (or response) functions or they do not (or their ability to contribute to a response is considerably at risk). The highest-level possibilities are conceptually clear; the underlying complexity matters in the playing out of the cycles of adaptation and counteradaptation and in potentially obfuscating the line between overall being held at risk or not. That is, there is a gray space of uncertainty where either (or both) of the principals is unsure whether forward-deployed forces continue to be safe and effective.

Conclusion

These influence points represent core U.S. national interests and the strategic conditions and operational capabilities required to sustain them. Competition and conflict will generate other, more specific interests to be protected at particular places and times, but these will be instrumental to defending the enduring influence points described here.

There is no clear demarcation between the defense of these influence points and the actions that will be required to protect them. Just as the protection of any one influence point will require a carefully orchestrated set of actions, the complexity of the U.S. legal, political, and economic systems means that some individual actions will affect multiple influence points simultaneously. In addition, these interests must be continually monitored to ensure that adversary action is not undermining them. Given this complex-

[18] For a discussion of causal complexity, see Charles C. Ragin, *The Comparative Method: Moving Beyond Qualitative and Quantitative Strategies*, Oakland: University of California Press, 1987. For a discussion of partial set membership in otherwise categorical outcomes under causal complexity, see Charles C. Ragin, *Fuzzy-Set Social Science*, Chicago: University of Chicago Press, 2000.

ity, formulating information requirements to understand when an adversary may be taking actions against these influence points will require a great deal of research and a rigorous process for collecting, analyzing, and presenting accurate status reports on a continuing basis.

Finally, U.S. influence points exist in a global ecosystem; there is no clear demarcation between domestic and foreign interests, just as there is no arbitrary separation of interests between the United States and its allies and partners. Indeed, one of the distinguishing asymmetries between the United States and the People's Republic of China is that most U.S. international interests are non-zero-sum, in that increasing security, liberty, and prosperity overseas enhances those same interests in the United States. Chinese international interests, on the other hand, are mostly zero-sum, where Chinese political, economic, or military gains are at the expense of adversary, or even client, states. This asymmetry suggests untapped potential in protecting and advancing U.S. interests through increased international cooperation.

CHAPTER SEVEN

Conclusion and Recommendations

We developed an alternative concept to address the challenge of competing with China and preparing for potential conflict in the cyber domain: selective overmatch. We developed this concept to examine operational postures based on two variables: whether the operational posture was in competition or in conflict and whether the effects sought were minimalist (lower risk, so less provocative and escalatory) or maximalist (higher risk and more aggressive and lethal). Arranging these two variables gave us four quadrants representing four different operational postures in cyberspace, each with its own set of capabilities and limitations. We identified key influence points for China and the United States and the actions that might be taken against them.

Findings and Insights

CYBERCOM and the CMF are operating at a high tempo, with teams fully engaged in planning and conducting operations, and little institutional overhead to support the reset and reconstitution of people and forces. Cyberspace operations forces are highly skilled personnel and require extensive training to fill critical roles on CMF teams. Often the pipeline for producing trained personnel struggles to sustain the pace and, as the U.S. Government Accountability Office (GAO) noted in a December 2022 report, service retention rates have also suffered. The nature of cyberspace operations, which require extensive ISR of potential targets due to shifting cyber terrain and the development of perishable capabilities, also stresses the need for capacity to meet the demands on cyber forces.

CYBERCOM has operated under a strategy of "defend forward" and the domain-driven operational concept of "persistent engagement" for several years, recognizing that a passive posture was not deterring adversaries or protecting the United States. In other words, the rope-a-dope strategy does not work if the adversary never gets tired and can continue to throw punches. But these concepts, while useful and necessary, appear in their current form to have some, but limited, success. Addressing challenges from revisionist powers requires extending these concepts and bringing more capabilities to bear that integrate across domains. Therefore, a new and complementary operational concept is required to identify categories of targets to be attacked or defended, a simple but comprehensive way to organize and visualize cyber operations against them, and a set of "meta-capabilities" to consider for more complete strategic thinking and operational design in cyberspace.

Selective overmatch provides this concept. It is a framework for evaluating what should be done against specific priority influence points, both offensively and defensively, in order to employ existing forces most effectively in the near term and to develop the highest-payoff new capabilities in the medium term. As CYBERCOM's operational posture vis-à-vis an adversary changes, so would the constellation of appropriate actions to be taken against that adversary's influence points at the policy, strategic, and operational levels. CYBERCOM does not currently have all the capabilities, force structure, operational relationships, and plans it would need to prosecute, or protect, the categories of influence points we identified using the framework.[1]

Each of these categories of meta-capabilities would offer more options—and more effective options—against each influence point in each quadrant. Properly employed, it would allow CYBERCOM to realistically assess which missions it should pursue and which missions should be delayed, transferred, or abandoned. This clearer, broader view of cyberspace, and a more

[1] We acknowledge that this is the case currently, but the array of targets and potential operations that CCMDs nominate through the cyber force mission alignment process indicates there is not a holistic view of how to best determine the application of cyberspace capabilities to achieve the most critical effects; rather, CYBERCOM must adjudicate across these competing demands and balance national-level and CCMD-level needs.

complete set of tools with which to operate, would give the strategic decisionmakers and operational commanders a clearer and more nuanced basis for force employment in the present and force development in the future. At the same time, developing the institutional capacity, capabilities, and relationships to carry out selective overmatch will require potentially significant effort in the coming years.

Selective overmatch and the concomitant I2E2 capabilities, if pursued, will require a fundamental reexamination of cyber force structure, operational relationships, and planning. It may also require changes in policy at the departmental level. As a concept, however, selective overmatch still requires further testing and validation before significant changes are pursued.

From a military service perspective, changes may be required in how they organize, train, equip, and present cyberspace forces as well as other forces, such as electronic warfare and information warfare. Although the services had developed cyberspace forces before the advent of the CMF, the current model is unusual in comparison to other types of forces. The CMF was "born joint"—that is, CYBERCOM defined the requirements, and the services each took on responsibility to meet those requirements. This is different from a more traditional model where the services organize, train, and equip forces for their operational domain that are then presented to CCMDs. Even special operations forces, despite the authorities that U.S. Special Operations Command has to influence requirements, are distinct across the services—a Navy SEAL is not the same as an Army Ranger, for example.

Recommendations

Near Term

Selective overmatch and the I2E2 set of meta-capabilities are a hypothesis designed to address a need for a broader and more comprehensive way to confront the challenge of revisionist powers like China. The concept has potentially far-reaching implications for doctrine, plans, organizational relationships, force generation and employment, and capabilities development as we have noted. Before advancing to these stages, however, CYBERCOM will need to test the hypothesis to determine how well it meets the stated need and develop implementation plans.

Recommendation 1: CYBERCOM should test the concept in a series of wargames and tabletop exercises to evaluate the degree to which it meets strategic and operational objectives. A wargame cannot "prove" whether a concept works or not given the necessarily artificial nature of the game, but it can help elucidate the requirements, identify risks, and aid in refining the concept. A wargame or series of wargames should include the regional CCMD and other experts who can speak knowledgeably to regional and functional issues, including and beyond cyberspace, as the purpose is to expand from a singularly domain-centric view to an integrated one. This recommendation can be executed in fiscal year 2023.

Medium Term

After a wargame series, experimentation with existing capabilities and teams can demonstrate how to make the concept "real" in an operational environment. This will require making choices to balance current operational demands with future force preparation, including having elements of the CMF participate in the experiments.

Recommendation 2: Using units from the Cyber National Mission Force, and in partnership with the services, CYBERCOM should develop an experimentation plan. The experiments can take a "crawl, walk, run" approach to the I2E2 meta-capabilities, starting with relatively simple approaches, such as an integrated cyber experiment that brings together traditional cyberspace forces, EW, and other information environment functions to formulate, test, and evaluate different mixes of specialized personnel, sensors, weapons, accesses, and inter-team coordination in order to prosecute Chinese influence points and protect U.S. influence points. The goal of this midterm action is to develop a replicable model for the creation of additional integrated cyber teams to bring all these enhanced capabilities to bear in all four cyber operational postures against all assigned influence points.

Further experiments would then address other aspects of the I2E2 meta-capabilities. For example, CYBERCOM could select elements from an NMT and a national CPT to develop integral cyber techniques, tactics, and procedures to offensively enable and defensively protect advanced kinetic weapon systems and platforms. Priority systems might include platforms that are likely to be critical to the early stages of a conflict, such as the F-35 aircraft, the B-21 stealth bomber, advanced destroyers (e.g., the Flight III DDG-51 when it is available), submarines, and hypersonic weapons. The teams would work actively with service and INDOPACOM plans and operations personnel to identify defensive threats to be countered and offensive opportunities to be exploited. The goal of this operational test and development effort will be to identify tactics, techniques, and procedures (TTPs), which may be employed at scale during any conflict in the INDOPACOM AOR.

These experiments can demonstrate the effectiveness of new team constructs, operational relationships, and existing capabilities. They can also feed directly into changes in operational plans and unit tactics, techniques and procedures to implement within existing force structure, and command relationships. At the same time, experimenters can work to identify desired but not yet available capabilities to inform research and development programs. An experimentation program of this type will be a multiyear effort that can begin in fiscal year 2024.

Long Term

Once CYBERCOM and other components of the DoD have tested the concept and meta-capabilities and determined that they make sense in an operational setting, adopting and employing them at scale will require changes across the DOTMLPF spectrum.

Recommendation 3: Contingent on a determination that the concept and meta-capabilities are necessary and effective, CYBERCOM should work with the other CCMDs, the services, the Joint Staff, and the Office of the Secretary of Defense to develop requirements for a revised force structure to organize, train, equip, and present integrated cyberspace forces to CYBERCOM.

Recommendation 4: Operational planning at CCMDs would benefit from better integration of cyberspace expertise from the start of the planning process. Although cyberspace operations are included (typically in the plan's operations annex as an appendix), the planning process still primarily focuses on operations in the physical domains as the primary efforts and operations in other domains (cyber, information) as secondary or supporting efforts. This integration will require CYBERCOM to work with the Joint Staff to revise the approach to planning, which often addresses domains in distinct silos.

Recommendation 5: This prior set of recommended changes will call for revisions in operational relationships, not only at the CCMD-to-CCMD level, as CYBERCOM may need operational control over a more diverse set of forces, but also at subordinate levels. These changes will need to be explored and refined as part of the roadmap from wargame to experiments to implementation. Changes to force allocation and assignment will require Secretary of Defense approval and, should the mission set of CYBERCOM expand to include additional cyber-adjacent capabilities, may also require an update to the Unified Command Plan and approval by the Commander in Chief.

Concluding Thoughts

Selective overmatch can provide CYBERCOM, the Joint Staff, and the Office of the Secretary of Defense an opportunity to comprehensively reevaluate current cyber operations in light of current competition and possible conflict with China. It provides both a framework for force employment, force development, and force design, as well as a means of conceptually organizing strategic thinking and operational design on a continuing basis. If accepted and implemented, it will allow for a clearer, more detailed view of a broader and more complex set of cyber variables, allowing for more deliberate risk management and operational decisionmaking.

Abbreviations

A2/AD	anti-access/area denial
AI	artificial intelligence
AOR	area of responsibility
BRI	China's Belt and Road Initiative
C5ISRT	command, control, communication, computer, cyber, intelligence, surveillance, reconnaissance, targeting
CCDR	combatant commander
CCG	Chinese Coast Guard
CCMD	combatant command
CCMF	Cyber Combat Mission Force
CCP	Chinese Communist Party
CCPPD	CCP Propaganda Department
CMC	Central Military Commission
CMF	Cyber Mission Force
CMT	cyber (combat) mission team
CNMF	Cyber National Mission Force
COA	course of action
CO-IPEs	cyber operations-integrated planning elements
CPF	Cyber Protection Force
CPT	cyber protection team
CYBERCOM	U.S. Cyber Command
DCO	defensive cyberspace operation
DoD	Department of Defense
DoDIN	Department of Defense Information Network
EW	electronic warfare
GDP	gross domestic product
GFW	Great Firewall
I2E2	integrated, integral, extended, expanded
ICBM	intercontinental ballistic missile

INDOPACOM	United States Indo-Pacific Command
ISR	intelligence, surveillance, and reconnaissance
IT	information technology
JCWA	Joint Cyber Warfighting Architecture
MC4	minimalist competition, maximalist competition, minimalist conflict, and maximalist conflict
NMT	national mission team
PAFMM	People's Armed Forces Maritime Militia
PAP	People's Armed Police
PBSC	Politburo Standing Committee
PLA	People's Liberation Army
PLAA	PLA Army
PLAAF	PLA Air Force
PLAN	PLA Navy
PLARF	PLA Rocket Force
PLASSF	PLA Strategic Support Force
PRC	People's Republic of China
RDT&E	research, development, test, and engineering
SCIO	State Council Information Office
SIIO	State Internet Information Office
TC	theater command
VPN	virtual private network
WMD	weapon of mass destruction

Bibliography

Air Force Materiel Command, "Air Force Announces Vanguard PEOs," *AFRL*, February 26, 2020. As of January 6, 2023:
https://www.afrl.af.mil/News/Article/2338745/air-force-announces-vanguard-peos/

Alexander, Keith B., "Statement of General Keith B. Alexander Commander United States Cyber Command Before the House Committee on Armed Services Subcommittee on Emerging Threats and Capabilities," March 16, 2011.

"Annual Briefing on Relationship Between National Security Agency and United States Cyber Command."

Bateman, Jon, *U.S.-China Technological "Decoupling": A Strategy and Policy Framework*, Washington, D.C.: Carnegie Endowment for International Peace, April 25, 2022. As of April 19, 2023:
https://carnegieendowment.org/2022/04/25/u.s.-china-technological-decoupling-strategy-and-policy-framework-pub-86897

Beauchamp-Mustafaga, Nathan, "Cognitive Domain Operations: The PLA's New Holistic Concept for Influence Operations," *Jamestown Foundation China Brief*, Vol. 19, No. 16, September 6, 2019. As of April 19, 2023:
https://jamestown.org/program/cognitive-domain-operations-the-plas-new-holistic-concept-for-influence-operations/

Bennett, Andrew, and Colin Elman, "Complex Causal Relations and Case Study Methods: The Example of Path Dependence," *Political Analysis*, Vol. 14, No. 3, Summer 2006, pp. 250–267.

Biden, Joseph R., *National Security Strategy*, Washington, D.C.: White House, October 2022.

Biden, Joseph R., *National Cybersecurity Strategy*, Washington, D.C.: White House, March 2023.

Brady, Anne-Marie, *Marketing Dictatorship: Propaganda and Thought Work in Contemporary China*, Lanham, Md.: Rowman & Littlefield, 2008.

Brady, Anne-Marie, "China's Foreign Propaganda Machine," Wilson Center, October 26, 2015. As of January 25, 2023:
https://www.wilsoncenter.org/article/chinas-foreign-propaganda-machine

Bureau of Economic Activity, "Gross Domestic Product." As of May 5, 2023:
https://www.bea.gov/data/gdp/gross-domestic-product#:~:text=In%20the%20fourth%20quarter%20of,a%20decrease%20in%20inventory%20investment

Burke, Edmund J., and Arthur Chan, "Chapter 6: Coming to a (New) Theater Near You: Command, Control, and Forces," in Phillip C. Saunders, Arthur S. Ding, Andrew Scobell, Andrew N. D. Yang, and Joel Wuthnow, eds., *Chairman Xi Remakes the PLA: Assessing Chinese Military Reforms*, Washington, D.C.: National Defense University Press, 2019. As of April 19, 2023:
https://ndupress.ndu.edu/Media/News/News-Article-View/Article/1747574/coming-to-a-new-theater-near-you-command-control-and-forces/

"CACI Awarded $51 Million Task Order to Provide Software Development Support to U.S. Air Force 90th Cyberspace Operations Squadron," press release, August 30, 2017. As of January 6, 2023:
https://www.businesswire.com/news/home/20170830005232/en/CACI-Awarded-51-Million-Task-Order-to-Provide-Software-Development-Support-to-U.S.-Air-Force-90th-Cyberspace-Operations-Squadron

Cheng, Dean, "Converting the Potential to the Actual: Chinese Mobilization Policies and Planning," in Andrew Scobell, Arthur S. Ding, Phillip C. Saunders, and Scott W. Harold, eds., *The People's Liberation Army and Contingency Planning in China*, Washington, D.C.: National Defense University Press, 2015. As of April 19, 2023:
https://ndupress.ndu.edu/Portals/68/Documents/Books/PLA-contingency/PLA-Contingency-Planning-China.pdf

"China Builds a Self-Repressing Society," *The Economist*, May 14, 2022. As of April 19, 2023:
https://www.economist.com/china/2022/05/14/china-builds-a-self-repressing-society

"China GDP Annual Growth Rate," *Trading Economics*, undated. As of April 19, 2023:
https://tradingeconomics.com/china/gdp-growth-annual

"China's Xi Bets Economic Growth Will Offset Misery from Covid," *Bloomberg News*, December 22, 2022. As of April 19, 2023:
https://www.bloomberg.com/news/articles/2022-12-22/china-s-xi-bets-economic-growth-will-offset-misery-from-covid

Cooperative Cyber Defense Center of Excellence (CCDCOE), Operation Orchard/Outside the Box (2007), *Cyber Law Toolkit*, September 6, 2007. As of April 19, 2023:
https://cyberlaw.ccdcoe.org/wiki/Operation_Orchard/Outside_the_Box_(2007)

Copeland, Dale C., "Economic Interdependence and War: A Theory of Trade Expectations," *International Security*, Vol. 20, No. 4, Spring 1996, pp. 5–41.

Couture, Mario, "Complexity and Chaos—State-of-the-Art: Overview of Theoretical Concepts," technical memorandum, Defence R&D Canada, August 2007.

Craig, Anthony, and Brandon Valeriano, "Realism and Cyber Conflict: Security in the Digital Age," in Davide Orsi, J. R. Avgustin, and Max Nurnus, eds., *Realism in Practice: An Appraisal*, Bristol: E-International Relations Publishing, 2018.

Cyber National Mission Force Public Affairs, "The Evolution of Cyber: Newest Subordinate Unified Command Is Nation's Joint Cyber Force," press release, December 19, 2022. As of January 11, 2023:
https://www.cybercom.mil/Media/News/Article/3250075/the-evolution-of-cyber-newest-subordinate-unified-command-is-nations-joint-cyber/

Cybersecurity and Infrastructure Security Agency, "National Critical Functions Set," website, undated. As of January 3, 2023:
https://www.cisa.gov/national-critical-functions-set

Davidson, Helen, "China Brings in 'Emergency' Level Censorship over Zero Covid Protests," *The Guardian*, December 2, 2022. As of April 19, 2023:
https://www.theguardian.com/world/2022/dec/02/china-brings-in-emergency-level-censorship-over-zero-covid-protests

"Defense FY 2023 IT and Cyberspace Activities Budget Request Highlights," June 9, 2022. As of December 29, 2022:
https://iq.govwin.com/neo/marketAnalysis/view/Defense-FY-2023-IT-and-Cyberspace-Activities-Budget-Request-Highlights/6638?researchTypeId=1&researchMarket=

Defense Innovation Unit, "Accelerating Commercial Technology for National Security," website, undated. As of December 29, 2022:
https://www.diu.mil

"Defense Primer: Cyberspace Operations," Congressional Research Service, IF10537, December 9, 2022. As of January 6, 2023:
https://crsreports.congress.gov/product/pdf/IF/IF10537

Di Pane, James, "Cyber Warfare and U.S. Cyber Command," *2023 Index of Military Strength*, Heritage Foundation, October 18, 2022. As of April 19, 2023:
https://www.heritage.org/military-strength/assessment-us-military-power/cyber-warfare-and-us-cyber-command

DoD—*See* U.S. Department of Defense.

Echevarria, Antulio J., "Clausewitz's Center of Gravity: It's Not What We Thought," *Naval War College Review*, Vol. 56, No. 1, Winter 2003, pp. 108–123.

Ellsworth, Robert, Andrew Goodpaster, and Rita Hauser, *America's National Interests: A Report from the Commission on America's National Interests*, Cambridge, Mass.: Belfer Center for Science and International Affairs, Harvard Kennedy School, July 1996.

Ellsworth, Robert, Andrew Goodpaster, and Rita Hauser, *America's National Interests: A Report from the Commission on America's National Interests*, Cambridge, Mass.: Belfer Center for Science and International Affairs, Harvard Kennedy School, July 2000.

Eriksson, Johan, and Giampiero Giacomello, "The Information Revolution, Security, and International Relations: (IR)relevant Theory?" *International Political Science Review*, Vol. 27, No. 3, July 2006, pp. 221–244.

Executive Order No. 14028, "Improving the Nation's Cybersecurity," May 12, 2021. As of January 20, 2023:
https://www.whitehouse.gov/briefing-room/presidential-actions/2021/05/12/executive-order-on-improving-the-nations-cybersecurity/

Feng, Coco, "Chinese Government Takes Minority Stake, Board Seat in TikTok Owner ByteDance's Main Domestic Subsidiary," *South China Morning Post*, August 17, 2021. As of January 5, 2023:
https://www.scmp.com/tech/big-tech/article/3145362/chinese-government-takes-minority-stake-board-seat-tiktok-owner

Feng, Yibing, "Beijing Banner Protest Ripples Outward as China Maintains Silence," *Voice of America*, October 20, 2022. As of April 19, 2023:
https://www.voanews.com/a/beijing-banner-protest-ripples-outward-as-china-maintains-silence/6799244.html

Finnemore, Martha, and Duncan B. Hollis, "Constructing Norms for Global Cybersecurity," *The American Journal of International Law*, Vol. 110, No. 3, July 2016, pp. 425–479.

Fiscal Year 2021 Budget Request for U.S. Cyber Command and Operations in Cyberspace, Hearing Before the House Committee on Armed Services, Subcommittee on Intelligence, Emerging Threats, and Capabilities, 116th Congress, March 4, 2020, Statement of Kenneth Rapuano, Assistant Secretary of Defense for Homeland Defense and Global Security and Principal Cyber Advisor.

Fischerkeller, Michael P., Emily O. Goldman, and Richard J. Harknett, *Cyber Persistence Theory: Redefining National Security in Cyberspace*, New York: Oxford University Press, 2022.

Garafola, Cristina L., Timothy R. Heath, Christian Curriden, Meagan L. Smith, Derek Grossman, Nathan Chandler, and Stephen Watts, *The People's Liberation Army's Search for Overseas Basing and Access*, Santa Monica, Calif.: RAND Corporation, RR-A1496-2, 2022. As of April 19, 2023:
https://www.rand.org/pubs/research_reports/RRA1496-2.html

Grossman, Derek, Christian Curriden, Logan Ma, Lindsey Polley, J. D. Williams, and Cortez A. Cooper III, *Chinese Views of Big Data Analytics*, Santa Monica, Calif.: RAND Corporation, RR-A176-1, 2020. As of April 19, 2023:
https://www.rand.org/pubs/research_reports/RRA176-1.html

Grossman, Derek, and Logan Ma, "A Short History of China's Fishing Militia and What It May Tell Us," The RAND Blog, April 6, 2020. As of April 19, 2023: https://www.rand.org/blog/2020/04/a-short-history-of-chinas-fishing-militia-and-what.html

Grotto, Andrew, "Deconstructing Cyber Attribution: A Proposed Framework and Lexicon," *IEEE Security & Privacy*, Vol. 18, No. 1, January–February 2020, pp. 12–20.

Heath, Timothy R., Derek Grossman, and Asha Clark, *China's Quest for Global Primacy: An Analysis of Chinese International and Defense Strategies to Outcompete the United States*, Santa Monica, Calif.: RAND Corporation, RR-A447-1, 2021, pp. 63–64. As of April 19, 2023: https://www.rand.org/pubs/research_reports/RRA447-1.html

Heflin, Judy, "The Single Greatest Educational Effort in Human History," *Language Magazine*, undated. As of January 6, 2023: https://www.languagemagazine.com/the-single-greatest-educational-effort-in-human-history/

Hodgson, Quentin E., Charles A. Goldman, Jim Mignano, and Karishma R. Mehta, *Educating for Evolving Operational Domains: Cyber and Information Education in the Department of Defense and the Role of the College of Information and Cyberspace*, Santa Monica, Calif.: RAND Corporation, RR-A1548-1, 2022. As of May 23, 2023: https://www.rand.org/pubs/research_reports/RRA1548-1.html

Hopf, Ted, "The Promise of Constructivism in International Relations Theory," *International Security*, Vol. 23, No. 1, Summer 1998, pp. 171–200.

Huntington, Samuel P., "The Erosion of American National Interests," *Foreign Affairs*, Vol. 76, No. 5, September–October 1997.

Inagaki, Kana, Leo Lewis, Ryan McMorrow, and Tom Mitchell, "Alibaba Founder Jack Ma Living in Tokyo Since China's Tech Crackdown," *Financial Times*, November 29, 2022. As of April 19, 2023: https://www.ft.com/content/2f7c7a10-2df3-4f1b-8d2a-eea0e0548713

Isnarti, Rika, "A Comparison of Neorealism, Liberalism, and Constructivism in Analysing Cyber War," *Andalas Journal of International Studies*, Vol. 5, No. 2, November 2016.

Jamison, Benjamin C., "A Constructivist Approach to a Rising China," *Journal of Indo-Pacific Affairs*, May 19, 2021.

Jervis, Robert, "Realism in the Study of World Politics," *International Organization*, Vol. 52, No. 4, Autumn 1998, pp. 971–991.

Jun Yan, Lianyong Feng, Artem Denisov, Alina Steblyanskaya, and Jan-Pieter Oosterom, "Complexity Theory for the Modern Chinese Economy from an Information Entropy Perspective: Modeling of Economic Efficiency and Growth Potential," *PLoS ONE*, Vol. 15, No. 1, 2020.

Keohane, Robert O., "The Demand for International Regimes, *International Organization*, Vol. 26, No. 2, Spring 1982, pp. 325–355.

Keohane, Robert O., "International Institutions: Can Interdependence Work?" *Foreign Policy*, Spring 1998, pp. 82–194.

Keohane, Robert O., and Joseph S. Nye, Jr., *Power and Interdependence*, 4th ed., New York: Pearson Longman, 2012.

Kokoshin, Andrei A., "2015 Military Reform in the People's Republic of China: Defense, Foreign, and Domestic Policy Issues," Belfer Center Paper, Belfer Center for Science and International Affairs, October 2016. As of April 19, 2023: https://www.belfercenter.org/sites/default/files/legacy/files/Military%20Reform%20China%20-%20web2.pdf

Lake, Anthony, "Defining the National Interest," *Proceedings of the Academy of Political Science*, Vol. 34, No. 2, 1981.

Lechner, Silviya, "Anarchy in International Relations," *International Studies*, 2017.

Lopez, C. Todd, "Cyber Command Expects Lessons from 2018 Midterms to Apply in 2020," *Defense.gov*, February 14, 2019. As of April 19, 2023: https://www.defense.gov/News/News-Stories/Article/Article/1758488/cyber-command-expects-lessons-from-2018-midterms-to-apply-in-2020

Lowther, Adam, and Casey Lucious, "A Call for Action: Defining the U.S. National Interest," *Atlantisch Perspectief*, Vol. 7, No. 37, 2013.

"Luo Shugang Was Appointed as the First Minister of Culture and Tourism" (*雒树刚被任命为首位文化和旅游部部长*), *Economic Daily-China Economic Net*, March 19, 2018. As of January 5, 2023: http://www.ce.cn/culture/gd/201803/19/t20180319_28524714.shtml

Ma Xiu, "PLA Rocket Force Organization: Executive Summary," China Aerospace Studies Institute, November 29, 2021. As of April 19, 2023: https://www.airuniversity.af.edu/CASI/Display/Article/2849664/pla-rocket-force-organization-executive-summary

Maizland, Lindsay, "China's Repression of Uyghurs in Xinjiang," Council on Foreign Relations, September 22, 2022. As of April 19, 2023: https://www.cfr.org/backgrounder/china-xinjiang-uyghurs-muslims-repression-genocide-human-rights

Maizland, Lindsay, and Eleanor Albert, "The Chinese Communist Party," Council on Foreign Relations, October 6, 2022. As of April 19, 2023: https://www.cfr.org/backgrounder/chinese-communist-party

Manjikian, Mary McEvoy, "From Global Village to Virtual Battlespace: The Colonizing of the Internet and the Extension of Realpolitik," *International Studies Quarterly*, Vol. 54, No. 2, June 2010, pp. 381–401.

Masters, John, and Will Merrow, "How Much Aid Has the U.S. Sent Ukraine? Here Are Six Charts," Council on Foreign Relations, December 16, 2022. As of January 3, 2023:
https://www.cfr.org/article/how-much-aid-has-us-sent-ukraine-here-are-six-charts

Mazarr, Michael J., Samuel Charap, Abigail Casey, Irina A. Chindea, Christian Curriden, Alyssa Demus, Bryan Frederick, Arthur Chan, John P. Godges, Eugeniu Han, et al., *Stabilizing Great-Power Rivalries*, Santa Monica, Calif.: RAND Corporation, RRA456-1, 2021. As of May 23, 2023:
https://www.rand.org/pubs/research_reports/RRA456-1.html

Mearsheimer, John J., "Anarchy and the Struggle for Power," in *The Tragedy of Great Power Politics*, New York: W. W. Norton, 2001.

"Meet China's New Politburo Standing Committee," Mercator Institute for China Studies, November 17, 2022. As of April 19, 2023:
https://merics.org/en/short-analysis/meet-chinas-new-politburo-standing-committee

Michelman-Ribeiro, Ariel, "A Primer on the Cyber Mission Force," Center for Naval Analysis (CNA), DIM-2022-U-033387-Final, September 2022.

Ministry of Foreign Affairs, "Full Text of the Report to the 20th National Congress of the Communist Party of China," October 25, 2022. As of April 19, 2023:
https://www.fmprc.gov.cn/eng/zxxx_662805/202210/t20221025_10791908.html

"More Than Half of Chinese Can Speak Mandarin," Xinhua News Agency, March 7, 2007. This article has since been removed by the Chinese government.

Morgan, David, "U.S. Says Chinese Vessels Harassed Navy Ship," *Reuters*, March 9, 2009. As of January 11, 2023:
https://www.reuters.com/article/us-usa-china-navy/u-s-says-chinese-vessels-harassed-navy-ship-idUSTRE52845A20090309

Morgenthau, Hans J., "The Policy of the USA," *Political Quarterly*, Vol. 22, No.1, 1951.

Morgenthau, Hans J., "What Is the National Interest of the United States?" *Annals of the American Academy of Political and Social Science*, Vol. 282, July 1952.

Morgenthau, Hans J., "Another 'Great Debate': The National Interest of the United States," *American Political Science Review*, Vol. 46, No. 4, December 1952.

Morgenthau, Hans J., *Politics Among Nations: The Struggle for Power and Peace*, 2nd ed., New York: Knopf, 1956.

Morris, Lyle, "What China's New Central Military Commission Tells Us About Xi's Military Strategy," Asia Society Policy Institute, October 27, 2022. As of April 19, 2023:
https://asiasociety.org/policy-institute/what-chinas-new-central-military-commission-tells-us-about-xis-military-strategy

Nakasone, Paul (Gen.), "CYBERCOM and NSA Chief: Cybersecurity Is a Team Sport," *Defense News*, December 6, 2021. As of December 29, 2022:
https://www.defensenews.com/outlook/2021/12/06/cybercom-and-nsa-chief-cybersecurity-is-a-team-sport

Nakasone, Paul M., "Posture Statement of Gen. Paul M. Nakasone, Commander, U.S. Cyber Command Before the 117th Congress," U.S. Cyber Command, press release, April 5, 2022. As of December 20, 2022:
https://www.cybercom.mil/Media/News/Article/2989087/posture-statement-of-gen-paul-m-nakasone-commander-us-cyber-command-before-the/

Nathan, Andrew, Bill Bishop, David Wertime, and Taisu Zhang, "China in the Panama Papers," ChinaFile Conversation, April 6, 2016. As of April 19, 2023:
https://www.chinafile.com/conversation/china-panama-papers

National Defense Authorization Act for Fiscal Year 2023, H.R. 7900, 117th Congress (2021–2022), *Congress.gov*, October 11, 2022. As of April 19, 2023:
https://www.congress.gov/bill/117th-congress/house-bill/7900

Navarre Chao, Lev, *Information Control Strategies of the Chinese Communist Party*, master's thesis, Harvard University Graduate School of Arts and Sciences, 2020.

Nye, Joseph S., Jr., "Redefining the National Interest," *Foreign Affairs*, Vol. 78, No. 4, July–August 1999.

Nye, Joseph S., Jr., *Cyber Power*, Cambridge, Mass: Belfer Center for Science and International Affairs, May 2010.

Nuechterlein, Donald E., "The Concept of National Interest," in *United States National Interests in a Changing World*, Louisville: University Press of Kentucky, 1973.

Ochmanek, David, *Restoring the Power Projection Capabilities of the U.S. Armed Forces*, Santa Monica, Calif.: RAND Corporation, CT-464, February 16, 2017. As of April 19, 2023:
https://www.rand.org/pubs/testimonies/CT464.html

Office of the Director of National Intelligence, *Annual Threat Assessment of the U.S. Intelligence Community*, February 6, 2023.

Office of the National Counterintelligence Executive, *Foreign Spies Stealing U.S. Economic Secrets in Cyberspace: Report to Congress on Foreign Economic Collection and Industrial Espionage, 2009–2011*, October 2011.

Office of the Under Secretary of Defense (Comptroller)/Chief Financial Officer, *Defense Budget Overview: United States Department of Defense Fiscal Year 2023 Budget Request*, April 2022.

Operational Test and Evaluation, Defense, "Exhibit R-2, RDT&E Budget Item Justification: PB 2021 Operational Test and Evaluation, Defense," February 2020. As of January 3, 2022:
https://apps.dtic.mil/descriptivesum/Y2021/Other/OTE/stamped/U_0605131OTE_6_PB_2021.pdf

Palmer, James, "Why China Is Cracking Down on Private Tutoring," *Foreign Policy*, July 28, 2021. As of January 5, 2023:
https://foreignpolicy.com/2021/07/28/china-private-tutoring-education-regulation-crackdown/

Pederson, Eric (Maj), MAJ Don Palermo, MAJ Stephen Fancey, and LCDR (Ret) Tim Blevins, "DoD Cyberspace: Establishing a Shared Understanding and How to Protect It," Air, Land, Sea, Space Application (ALSSA) Center, January 1, 2022. As of January 6, 2023:
https://www.alsa.mil/News/Article/2891794/dod-cyberspace-establishing-a-shared-understanding-and-how-to-protect-it/

PEO STRI, "Persistent Cyber Training Environment (PCTE)," website, undated. As of January 3, 2022:
https://www.peostri.army.mil/persistent-cyber-training-environment-pcte

Petallides, Constantine J., "Cyber Terrorism and IR Theory: Realism, Liberalism, and Constructivism in the New Security Threat," *Inquiries Journal/Student Pulse*, Vol. 4, No. 3, 2012.

Pew Research Center, "Public Trust in Government: 1958–2022," June 6, 2022. As of January 3, 2023:
https://www.pewresearch.org/politics/2022/06/06/public-trust-in-government-1958-2022/

Phillips, Tom, "All Mention of Panama Papers Banned from Chinese Websites," *The Guardian*, April 5, 2016. As of April 19, 2023:
https://www.theguardian.com/news/2016/apr/05/all-mention-of-panama-papers-banned-from-chinese-websites

Pollpeter, Kevin L., Michael S. Chase, and Eric Heginbotham, *The Creation of the PLA Strategic Support Force and Its Implications for Chinese Space Operations*, Santa Monica, Calif.: RAND Corporation, RR2058-AF, 2017. As of April 19, 2023:
https://www.rand.org/pubs/research_reports/RR2058.html

Pomerleau, Mark, "U.S. Military to Blend Electronic Warfare with Cyber Capabilities," *C4ISRNet*, April 14, 2021. As of January 4, 2022:
https://www.c4isrnet.com/electronic-warfare/2021/04/14/us-military-to-blend-electronic-warfare-with-cyber-capabilities/

Pomerleau, Mark, "Key Lawmakers in Favor of Keeping 'Dual Hat' Arrangement Between Cybercom and NSA," *DefenseScoop*, November 17, 2022. As of April 19, 2023: https://defensescoop.com/2022/11/17/two-key-lawmakers-in-favor-of-keeping-dual-hat-arrangement-between-cybercom-and-nsa/

Prince, Ryan, "Space Delta 6 Protects Space and Cyberspace," Peterson-Schriever Garrison Public Affairs, undated. As of December 29, 2022: https://www.schriever.spaceforce.mil/News/Article-Display/Article/2445974/space-delta-6-protects-space-and-cyberspace/

Ragin, Charles C., *The Comparative Method: Moving Beyond Qualitative and Quantitative Strategies*, Oakland: University of California Press, 1987.

Ragin, Charles C., *Fuzzy-Set Social Science*, Chicago: University of Chicago Press, 2000.

Rakhmat, Muhammad Zulfikar, "The Belt and Road Initiative in the Gulf: Building 'Oil Roads' to Prosperity," Middle East Institute, March 12, 2019. As of April 19, 2023: https://www.mei.edu/publications/belt-and-road-initiative-gulf-building-oil-roads-prosperity

Reardon, Robert, and Nazli Choucri, "The Role of Cyberspace in International Relations: A View of the Literature," paper prepared for the 2012 International Studies Association Annual Convention, San Diego, Calif., April 1, 2012.

Reddy, Rahul Karan, "China's Anti-Corruption Campaign: Tigers, Flies, and Everything in Between," *The Diplomat*, May 12, 2022. As of April 19, 2023: https://thediplomat.com/2022/05/chinas-anti-corruption-campaign-tigers-flies-and-everything-in-between/

Rhodes, Mary Lee, and Elizabeth Eppel, "Public Administration and Complexity—Or How to Teach Things We Can't Predict?" *Complexity, Governance & Networks*, Vol. 4, No. 1, January 2018.

Roberts, Margaret E., *Censored: Distraction and Diversion Inside China's Great Firewall*, Princeton, N.J.: Princeton University Press, 2018.

Rocha, Marcio, and Daniel Farias da Fonseca, "The Cyber Issue and Realist Thinking," *Revista da Escola de Guerra Naval (Rio de Janeiro)*, Vol. 25, No. 2, May–August 2019, pp. 517–543.

Roskin, Michael G., *National Interest: From Abstraction to Strategy*, Carlisle, Penn.: Strategic Studies Institute, U.S. Army War College, 1994.

Ruggie, John Gerard, "International Regimes, Transactions, and Change: Embedded Liberalism in the Postwar Economic Order," *International Organization*, Vol. 36, No. 2, Spring 1982, pp. 379–415.

Saich, Anthony, ed., *The Rise to Power of the Communist Party: Documents and Analysis*. Armonk, N.Y.: M. E. Sharpe, 1996.

Saich, Anthony, and David E. Apter, *Revolutionary Discourse in Mao's Republic*. Cambridge, Mass.: Harvard University Press, 1994.

Sakamoto, Shigeki, "China's New Coast Guard Law and Implications for Maritime Security in the East and South China Seas," *Lawfare*, February 16, 2021. As of April 19, 2023:
https://www.lawfareblog.com/chinas-new-coast-guard-law-and-implications-maritime-security-east-and-south-china-seas

Scobell, Andrew, Edmund J. Burke, Cortez A. Cooper III, Sale Lilly, Chad J. R. Ohlandt, Eric Warner, and J. D. Williams, *China's Grand Strategy: Trends, Trajectories, and Long-Term Competition*, Santa Monica, Calif.: RAND Corporation, RR-2798-A, 2020. As of May 7, 2023:
https://www.rand.org/pubs/research_reports/RR2798.html

Shats, Daniel, "Chinese Views of Effective Control: Theory and Action," China Aerospace Studies Institute, September 2022. As of April 19, 2023:
https://www.airuniversity.af.edu/Portals/10/CASI/documents/Research/Other-Topics/2022-10-03%20Effective%20Control.pdf

"Special Report 2021: China: Transnational Repression Origin Country Case Study," Freedom House, undated. As of April 19, 2023:
https://freedomhouse.org/report/transnational-repression/china

State Council Information Office, "China and the World in the New Era," white paper, September 27, 2019. As of April 19, 2023:
https://english.www.gov.cn/archive/whitepaper/201909/27/content_WS5d8d80f9c6d0bcf8c4c142ef.html

Steward, Dana, "Colors of Money," Defense Acquisition University, June 2020. As of June 6, 2023:
https://media.dau.edu/media/1_7aq33ev7

Tannenwald, Nina, "Stigmatizing the Bomb: Origins of the Nuclear Taboo," *International Security*, Vol. 29, No. 4, Spring 2005, pp. 5–49.

Taylor Fravel, M., *Active Defense: China's Military Strategy Since 1949*, Princeton, N.J.: Princeton University Press, 2019. As of April 19, 2023:
https://press.princeton.edu/books/hardcover/9780691152134/active-defense

Tiezzi, Shannon, "What Is the CPPCC Anyway?" *The Diplomat*, March 4, 2021. As of April 19, 2023:
https://thediplomat.com/2021/03/what-is-the-cppcc-anyway/

Trump, Donald J., *National Security Strategy of the United States of America*, Washington, D.C.: White House, December 2017.

Turner, John R., and Rose M. Baker, "Complexity Theory: An Overview with Potential Applications for the Social Sciences," *Systems*, Vol. 7, No. 1, 2019.

U.S. Cyber Command, "CYBER 101: Defend Forward and Persistent Engagement," website, October 25, 2022. As of January 4, 2022: https://www.cybercom.mil/Media/News/Article/3198878/cyber-101-defend-forward-and-persistent-engagement/

U.S. Cyber Command, "CYBER 101: Cyber Mission Force," website, November 1, 2022. As of December 29, 2022: https://www.cybercom.mil/Media/News/Article/3206393/cyber-101-cyber-mission-force/

U.S. Cyber Command, "CYBER 101: Sixteenth Air Force (AFCYBER)," website, November 22, 2022. As of December 29, 2022: https://www.cybercom.mil/Media/News/Article/3226434/cyber-101-sixteenth-air-force-afcyber/

U.S. Cyber Command, "CYBER 101: US Army Cyber Command (ARCYBER)," website, November 30, 2022. As of December 29, 2022: https://www.cybercom.mil/Media/News/Article/3232195/cyber-101-us-army-cyber-command-arcyber/

U.S. Cyber Command, "CYBER 101: US Coast Guard Cyber Command (CGCYBER)," website, December 15, 2022. As of December 29, 2022: https://www.cybercom.mil/Media/News/Article/3247521/cyber-101-us-coast-guard-cyber-command-cgcyber/

U.S. Cyber Command, "CYBER 101: US Fleet Cyber Command (FCC)," website, December 20, 2022. As of December 29, 2022: https://www.cybercom.mil/Media/News/Article/3251285/cyber-101-us-fleet-cyber-command-fcc/

U.S. Cyber Command, "CYBER101: US Marine Corps Forces Cyberspace Command (MARFORCYBER)," website, December 27, 2022. As of December 29, 2022: https://www.cybercom.mil/Media/News/Article/3254942/cyber101-us-marine-corps-forces-cyberspace-command-marforcyber/

U.S. Cyber Command, "U.S. Cyber Command 2022 Year in Review," website, December 30, 2022. As of January 6, 2023: https://www.cybercom.mil/Media/News/Article/3256645/us-cyber-command-2022-year-in-review/

U.S. Cyberspace Solarium Commission, "Final Report of the Cyberspace Solarium Commission," December 30, 2018. As of January 6, 2023: https://www.solarium.gov/report

U.S. Department of Defense, *Joint Publication 3-60, Joint Targeting*, January 31, 2013.

U.S. Department of Defense, "Cyber Mission Force Achieves Full Operational Capability," website, May 17, 2018. As of December 20, 2022: https://www.defense.gov/News/News-Stories/Article/Article/1524747/cyber-mission-force-achieves-full-operational-capability/

U.S. Department of Defense, *Joint Publication 3-12, Cyberspace Operations*, June 8, 2018.

U.S. Department of Defense, *Joint Doctrine Note 1-19, Competition Continuum*, June 3, 2019.

U.S. Department of Defense, *Joint Publication 5-0, Joint Planning*, December 1, 2020.

U.S. Department of Defense, *Dictionary of Military and Associated Terms*, Washington, D.C.: The Joint Staff, November 2021.

U.S. Department of Defense, "Department of Defense Software Modernization," memorandum for senior Pentagon leadership, February 1, 2022. As of December 29, 2022: https://media.defense.gov/2022/Feb/03/2002932833/-1/-1/1/DEPARTMENT-OF-DEFENSE-SOFTWARE-MODERNIZATION-STRATEGY.PDF

U.S. Department of Defense, "DoD Directive 5000.01: The Defense Acquisition System (Change 1)," July 28, 2022.

U.S. Department of Defense, "DoD 7000.14-R: Financial Management Regulation, Vol. 2B, Ch. 5," September 2022. As of December 29, 2022: https://comptroller.defense.gov/portals/45/documents/fmr/current/02b/02b_05.pdf

U.S. Department of Defense, *2022 National Defense Strategy of the United States of America*, October 27, 2022.

U.S. Department of Defense, *Military and Security Developments Involving the People's Republic of China*, Annual Report to Congress, November 29, 2022. As of April 19, 2023: https://media.defense.gov/2022/Nov/29/2003122279/-1/-1/1/2022-MILITARY-AND-SECURITY-DEVELOPMENTS-INVOLVING-THE-PEOPLES-REPUBLIC-OF-CHINA.PDF

U.S. Government Accountability Office, "Personnel and Support Needed for Joint Cyber Centers," DODIG-2015-048, December 8, 2014. As of January 6, 2023: https://media.defense.gov/2018/Aug/27/2001958644/-1/-1/1/DODIG-2015-048.PDF

U.S. Government Accountability Office, "Defense Acquisitions: Joint Cyber Warfighting Architecture Would Benefit from Defined Goals and Governance," GAO-21-68, November 19, 2020.

U.S. Government Accountability Office, "Defense Acquisitions: Cyber Command Needs to Develop Metrics to Assess Warfighting Capabilities," March 30, 2022. As of January 6, 2022: https://www.gao.gov/assets/gao-22-104695.pdf

U.S. Government Accountability Office, "Military Cyber Personnel: Opportunities Exist to Improve Service Obligation Guidance and Data Tracking," GAO-23-105423, December 21, 2022. As of January 6, 2023: https://www.gao.gov/products/gao-23-105423

Wall, Andru E., "Demystifying the Title 10-Title 50 Debate: Distinguishing Military Operations, Intelligence Activities and Covert Action," *Harvard National Security Journal*, Vol. 3, No. 1, 2011, pp. 85–142.

Waltz, Kenneth, *Man, The State, and War*, New York: Columbia University Press, 1959.

Wang Yi, Ministry of Foreign Affairs, "The Right Way for China and the United States to Get Along in the New Era," speech at the Asia Society, New York, September 22, 2022. As of April 19, 2023: https://www.fmprc.gov.cn/mfa_eng/wjb_663304/wjbz_663308/2461_663310/202209/t20220923_10770469.html

Warner, Michael, "U.S. Cyber Command's First Decade," Hoover Institution National Security, Technology, and Law Aegis Paper No. 2008, December 3, 2020.

Weeks, Michael R., "Chaos, Complexity and Conflict," *Air and Space Power Chronicles*, Vol. 16, 2001.

Wendt, Alexander, "Anarchy Is What States Make of It: The Social Construction of Power Politics," *International Organization*, Vol. 46, No. 2, Spring 1992, pp. 391–425.

Wessley, Brett, "Evolution of U.S. Cyber Operations and Information Warfare," *RealClear Defense*, June 10, 2017. As of March 14, 2023: https://www.realcleardefense.com/articles/2017/06/10/evolution_of_us_cyber_operations_and_information_warfare_111562-2.html

"What Does Xi Jinping's China Dream Mean?" *BBC News*, June 6, 2013. As of April 19, 2023: https://www.bbc.com/news/world-asia-china-22726375

Wilde, Gavin, and Jon Bateman, "Russia's Wartime Cyber Operations in Ukraine: Military Outcomes and Drivers," CYBERWARCON Conference, November 10, 2022. As of January 6, 2023: https://www.cyberwarcon.com

Williams, Heather J., Luke J. Matthews, Pauline Moore, Matthew A. DeNardo, James V. Marrone, Brian A. Jackson, William Marcellino, and Todd C. Helmus, *A Dangerous Web: Mapping Racially and Ethnically Motivated Violent Extremism*, Santa Monica, Calif.: RAND Corporation, RB-A1841-1, 2022. As of May 7, 2023:
https://www.rand.org/pubs/research_briefs/RBA1841-1.html

Wo-Lap Lam, Willy, "'Stability Maintenance' Gets a Major Boost at the National People's Congress," *Jamestown Foundation China Brief*, Vol. 19, No. 6, March 22, 2019. As of April 19, 2023:
https://jamestown.org/program/stability-maintenance-gets-a-major-boost-at-the-national-peoples-congress/

Wright, Teresa "Protests in China Are Not Rare—but the Current Unrest Is Significant," *The Conversation*, November 30, 2022. As of April 19, 2023:
https://theconversation.com/protests-in-china-are-not-rare-but-the-current-unrest-is-significant-195622

Yang, Stephanie, "As China Shuts Out the World, Internet Access from Abroad Gets Harder Too," *Los Angeles Times*, June 23, 2022. As of April 19, 2023:
https://www.latimes.com/world-nation/story/2022-06-23/china-great-firewall-foreign-domestic-virtual-censorship